IEE POWER ENGINEERING SERIES 5

SERIES EDITORS   DR. A.T. JOHNS
                 G. RATCLIFFE
                 PROF. A. WRIGHT

POWER SYSTEM ECONOMICS

Previous volumes in this series

Volume 1   Power circuit breaker theory & design
           C.H. Flurscheim (Editor)

Volume 2   Electric fuses
           A. Wright & P.G. Newberry

Volume 3   Z-transform electromagnetic transient
           analysis in high-voltage networks
           W. Derek Humpage

Volume 4   Industrial microwave heating
           A.C. Metaxes & R.J. Meredith

POWER SYSTEM ECONOMICS

T. W. Berrie

PETER PEREGRINUS LTD
on behalf of the
Institution of Electrical Engineers

Published by Peter Peregrinus Ltd., London, UK.

©1983: Peter Peregrinus Ltd

British Library Cataloguing in Publication Data

Berrie, T.W.
    Power system economics.—(IEE power engineering series; 5)
    1. Electric power production — Economic aspects
    2. Electric utilities — Economic aspects
    I. Title   II. Series
    621.31   TK 1001

ISBN 0-906048-88-5

Printed in England by Short Run Press Ltd., Exeter

Contents

Introduction   Themes and issues in power system economics

1	**Energy resources, sources and uses**	**1**
1.1	The energy background	1
	1.1.1 Introduction	1
	1.1.2 Non-renewable fuels	3
	1.1.3 Proved reserves of non-renewables	8
	1.1.4 Renewable but exhaustible fuels	14
	1.1.5 Inexhaustible fuels	14
	1.1.6 Special aspects for developing countries	16
1.2	The future world energy scene	17
	1.2.1 Introduction	17
	1.2.2 Substitution between dominant fuels	18
	1.2.3 World energy supply-demand position	19
	1.2.4 Coal production	20
	1.2.5 Petroleum production	22
	1.2.6 Natural-gas production	22
	1.2.7 Nuclear production	23
	1.2.8 Hydro production	24
	1.2.9 Renewable energy production	25
	1.2.10 Unconventional oil and gas	26
	1.2.11 Conservation	26
	1.2.12 Energy demand	26
	References	29
2	**The power system investment, reliability and pricing rules**	**31**
2.1	Introduction	31
	2.1.1 The power sector	31
	2.1.2 Power system economics	32
	2.1.3 Early history	32
	2.1.4 The broader view	32
	2.1.5 The subjects covered	33
2.2	Objectives of power system investment	33
	2.2.1 The basic rules	33
	2.2.2 Different viewpoints	35
	2.2.3 National economic viewpoint	35
	2.2.4 Increasing aggregate consumption	35
	2.2.5 Net present cost	36
	2.2.6 More equal distribution of income	37
	2.2.7 Advice on methods	37

2.3 Pricing objectives 38
   2.3.1 Introduction 38
   2.3.2 The pricing rule 38
   2.3.3 Marginal cost pricing 39
   2.3.4 Time-related tariffs 39
   2.3.5 Practical constraints on tariffs 40
   References 41

3 **Reliability standards** 43
3.1 The reliability rule 43
   3.1.1 Introduction 43
   3.1.2 Optimum reliability standard 43
   3.1.3 Cost incurred 44
   3.1.4 Outage costs 45
   3.1.5 A pragmatic approach 45
3.2 Practical approaches 47
   3.2.1 A practical methodology and model 47
   3.2.2 Feedbacks in the model 49
   3.2.3 Generating-plant reliability 51
   3.2.4 Measures of reliability 51
   3.2.5 Reliability indices 52
   3.2.6 Comparison of different reliability indices 52
3.3 How to calculate outage costs 53
   3.3.1 Introduction 53
   3.3.2 Direct and indirect outage costs 53
   3.3.3 Estimating outage costs 56
   3.3.4 Willingness-to-pay 57
   3.3.5 Effect on productive activity 57
   3.3.6 Effect of electricity outages on residential consumers 58
3.4 Appendix: System costs and outage costs: A formal derivation 60
   3.4.1 Keeping the price constant 60
   3.4.2 Allowing the price to vary 64
   References 66

4 **Load forecasting** 69
4.1 Identification of important problems 69
   4.1.1 Introduction 69
   4.1.2 Reasons for making load forecasts 69
   4.1.3 Load forecasting period 71
   4.1.4 Load forecasting principles 71
4.2 Load forecasting in practice 73
   4.2.1 Introduction 73
   4.2.2 The price of flexibility 73
   4.2.3 Use of statistics 73
   4.2.4 The database 75
4.3 The three basic methods of load forecasting 76
   4.3.1 Introduction 76
   4.3.2 Trend forecasting 77
   4.3.3 Market forecasting 77
   4.3.4 Macro-economic and econometric forecasts 80
   4.3.5 Uncertainty 80
   4.3.6 Economic load forecasts: Cost-benefit criteria 82
   References 84

5	**Fuels and generating plant mix**		**86**
	5.1	Fuels and generating plant	86
		5.1.1 Introduction	86
		5.1.2 Background plan	88
		5.1.3 Autogeneration and cogeneration	90
	5.2	Determining background plans	90
		5.2.1 Plant to load balances	90
		5.2.2 Basic situation	91
		5.2.3 Fuel cost savings	91
		5.2.4 Approximate optimum plant-mix in the background plan	93
		5.2.5 Marginal analysis	93
	5.3	Appendix: Load and plant duration curves	94
		5.3.1 Introduction	94
		5.3.2 Classification	97
		5.3.3 Construction of curves	99
		5.3.4 Some useful curves	100
		5.3.5 Acknowledgment	102
		References	102
6	**Energy sector modelling**		**104**
	6.1	Background to modelling	104
		6.1.1 Introduction	104
		6.1.2 Need for long-term demand projections	105
		6.1.3 National organisations making energy projections	106
	6.2	Recommended energy model structure	108
		6.2.1 Methodology	108
		6.2.2 Model details	110
	6.3	Energy model applications	113
		(a) Projection of national demand for energy	114
		(b) Energy balance in a nuclear based economy	114
		(c) National transport policy	114
		(d) International energy projections	114
		(e) Multi-national corporations and private companies	114
		References	115
7	**Power sector modelling**		**117**
	7.1	Basic approaches	117
		7.1.1 Introduction	117
		7.1.2 Basic approach to power sector modelling: Decomposition	118
		7.1.3 Power-system modelling	120
		7.1.4 Simplified example	121
	7.2	A worked example	122
		7.2.1 Introduction	122
		7.2.2 Importance of defining reliability standards	124
		7.2.3 Detailed example	125
		7.2.4 Summing up on generation planning	130
		7.2.5 Network limitations	131
	7.3	Network configuration synthesis	132
		7.3.1 Introduction	132
		7.3.2 Minimum circuit length criterion	132
		7.3.3 Group transfer criterion	133
		7.3.4 Perturbation methods of solution	135

		7.3.5	Summary	137
7.4		Appendix: Principles of power system economic modelling	137
		7.4.1	Introduction	137
		7.4.2	Marginal analysis	138
		7.4.3	Simulation models	139
		7.4.4	Global models	140
		7.4.5	Combined cost method	146
		7.4.6	Integer programming	147
		References	148

8	**Alternative generating projects**	**151**
8.1	Choice of methods	151
	8.1.1	Introduction	151
	8.1.2	Constant load-factor methods	152
	8.1.3	Net effective cost method	153
	8.1.4	Applicability of the net effective cost method	156
8.2	Smaller or rapidly expanding or mixed hydro-thermal systems	157
	8.2.1	Introduction	157
	8.2.2	Comparing alternative sequences of generaing plants	157
	8.2.3	Total present costs	159
8.3	Economic return	159
	8.3.1	Introduction	159
	8.3.2	Attributable benefits	161
	8.3.3	Acceptability of the economic return	161
8.4	Summary	162
8.5	Appendix: Comparison of alternative projects	162
	8.5.1	Introduction	162
	8.5.2	Sequence of development	163
	8.5.3	Period for present valuing	168
	8.5.4	Method of comparison	168
	8.5.5	Sensitivity analysis of cost parameters	174
	8.5.6	Sensitivity to shadow pricing	174
	8.5.7	Sensitivity to load growth	174
	References	175

9	**Pricing and load management**	**177**
9.1	Marginal-cost pricing	177
	9.1.1	Introduction	177
	9.1.2	Tariff components	177
	9.1.3	Capacity-related charges	178
	9.1.4	Energy-related charges	178
	9.1.5	Consumer-related charges	178
	9.1.6	Practical tariff-making	179
	9.1.7	Commercial contribution to tariff-making	179
	9.1.8	Accounting approach to tariff-making	180
	9.1.9	Economist approach to tariff-making	180
	9.1.10	Putting the contributions together: The overall strategy	180
9.2	Load management	181
	9.2.1	Introduction	181
	9.2.2	Control of peak and off-peak usage	181
	9.2.3	Load management by pricing	183
	9.2.4	Peak-load management by pricing	187
	9.2.5	Load control by direct means	187

		9.2.6	Centralised control	188
9.3		Interactive load control: Spot pricing		189
		9.3.1	Introduction	189
		9.3.2	Spot pricing	190
		9.3.3	The simple case	190
		9.3.4	Simple graphical solution to the problem	191
		9.3.5	Spot pricing and the investment rule	194
9.4		Summary		196
9.5		Appendix A: Calculating ideal long-run marginal costs		196
		9.5.1	Introduction	196
		9.5.2	Capacity-related costs	196
		9.5.3	LRMC capacity-related costs	197
		9.5.4	Marginal energy-related costs	199
		9.5.5	Consumer-related costs	200
9.6		Appendix B: Extract from world bank staff working paper 340		200
		9.6.1	Allocation of capacity and energy costs among peak and off-peak users	200
		References		205

10	Network economics			207
10.1	Introducing geography			207
		10.1.1	The Problem	207
		10.1.2	Multi-nodal systems	208
10.2	Functions of networks			208
		10.2.1	Introduction	208
		10.2.2	Planned transfer function of networks	209
		10.2.3	Capacity pooling function of networks	210
		10.2.4	Economic-operation function of networks	210
		10.2.5	Networks acting as multi-purpose projects	210
		10.2.6	Other functions of networks	211
10.3	Economic justification of a network			211
		10.3.1	Finding a least-cost solution	211
		10.3.2	Economic return	212
10.4	Transmission			212
		10.4.1	Introduction	212
		10.4.2	Major networks	218
		10.4.3	Likely developments of existing networks	219
10.5	Distribution			220
		10.5.1	Economic justification	220
		10.5.2	System reinforcement function	221
		10.5.3	System rehabilitation function	221
		10.5.4	System replacement function	222
		10.5.5	Distribution reliability	222
		10.5.6	Radial distribution systems	222
		10.5.7	Ring systems	225
		10.5.8	Continuous network development	226
		10.5.9	Economics of a distribution planning plant margin	227
		10.5.10	Need for more case studies	229
10.6	Summary			241
		References		242

| 11 | Rural electrification | | | 243 |
| 11.1 | Rural energy | | | 243 |

		11.1.1	Introduction	243
		11.1.2	Quantity and quality of rural energy	243
		11.1.3	Pattern of rural energy use	248
	11.2	Rural electrification		248
		11.2.1	Introduction	248
		11.2.2	Special characteristics of rural electrification	249
		11.2.3	Ground rules for rural electrification	251
	11.3	Appendix A: Rural electrification study: Work programme for economics		253
		11.3.1	Establish the marginal costs of supply	253
		11.3.2	Economics of connecting pumping loads, flour mills, motor drives etc.	253
		11.3.3	Economics of auto-generation	254
		11.3.4	Quality of supply	254
		11.3.5	Any expected development of tourism?	254
		11.3.6	Resource capabilities	255
		11.3.7	Calculation of economic benefits	255
		11.3.8	Existing and future likely tariff levels and structures	256
		11.3.9	Analysis of consumer survey	257
		11.3.10	Independent sales and demand forecasts	258
		11.3.11	Long-range load forecasts	258
		11.3.12	Methodology to be used for economic justification	259
		11.3.13	Fuel costs and energy policy	259
		11.3.14	Justifying particular projects in the development programme	260
		11.3.15	Capital budgets	260
	11.4	Appendix B: Economic return of a rural electrification project: A worked example		260
		11.4.1	Introduction	260
		11.4.2	Domestic and commercial consumers	261
		11.4.3	Irrigation and industrial consumers	263
		11.4.4	Sensitivity of rate of return to valuation of benefits	264
		References		269
12	Application to developing countries			270
	12.1	Background		270
		12.1.1	Introduction	270
		12.1.2	Main reasons for the difference	270
	12.2	Main considerations		272
		12.2.1	The economy	272
		12.2.2	The power sector	273
		12.2.3	Justification of projects	273
		12.2.4	Finance	274
		References		275
		Addendum 1: Risk analysis in project economics		276
		Addendum 2: In using discounting techniques in project analysis		289

Themes and issues in power system economics

This book has been written for a variety of reasons; in particular, it has given the author the opportunity after thirty years experience in power systems to explain his approach to the interlocking problems in power-system economics. The book is also an integrating paper following a symposium on power-system economics at Imperial College, London, UK, from 24th to 28th March 1980, to which the author owes some of the ideas and material. Again, it is a reference book for a wide spectrum of academics, planners, public utility operators and managers, government officials, consultants, and appraisal officers from national and international lending agencies. The author believes there should be one book on power-system economics to which all can turn for an across-the-board approach, integrating separate viewpoints and disciplines, e.g. economics, finance, engineering and management.

The book has two other important rôles: the first is to broadcast the latest developments in power-system economics, the second to put into simple language complex material written by economists, most publications on power-system economics still being written by economists. Many engineers find economists' publications incomprehensible, even though the engineer's tools of algebra and differential calculus are used.

The author has attempted a balanced use of his career between developed and developing countries, between generation and networks. He believes that power-system economics may be more important for developing than for developed countries; most developed countries have virtually constructed their power systems; no developing countries have reached this stage and some are in a 'green-field' situation. Again, all countries must make savings in investment and fuel, but shortage of these resources in most developing countries is critical. For the whole developing world capital investment of about US$450 billion is required for electricity in the decade 1981 to 1990[1] *. The investment needs for electricity in the United States for that period are about US$400 billion. Equally vast sums are needed for fuel costs.

All power utilities must be convinced they are getting full value from the vast sums spent. Having convinced themselves, the utilities must similarly convince their shareholders, governments and

* Throughout the book references are given at the end of each chapter

financiers. Perhaps the utility is at risk by expanding too rapidly? Shareholders and government might react similarly on this point since many utilities are government owned and all are government controlled. Perhaps the vast sums should be spent in other sectors of the economy: agriculture, transport, education and health are popular alternatives. One World Bank President is said to have had a chart in his office noting the number of hospitals, schools, buses, water pumps etc., which could be bought for the cost of just one power station. He knew the question which preoccupies governments: why spend so much in any one sector? One important job of power-system economics is to answer this question. The 'honeymoon' for electricity which started about 1900 is over.

All examples in the book are based on actual cases to illustrate the text and enable practitioners to work out their own problems after looking at the references.

Power-system economics is still changing in theory, practice and emphasis, partly due to the importance given to conserving energy and to cutting down on the use of conventional non-renewable fuels like coal, oil and gas, still the main fuels for power stations. Because of this aspect of change the author invites the reader to take an active rôle when reading both text and examples, adding a gloss from his own experience, thus constructing a vital bridge between reader and author.

Every effort has been made to trace all the copyright holders, but if any have been inadvertently overlooked the author apologises and will be pleased to make the necessary adjustment to the text at the first reasonable opportunity.

General Outline

The first seven chapters deal with power-system economics in its relation to world energy supply and demand, electricity planning standards, electricity demands, fuels and electricity production, together with the principles of energy sector and power-system modelling. The next two chapters deal with electricity generating station selection and pricing, giving special mention to fuel mix, long-term development programmes and project justification. The tenth chapter covers the changing structure of power networks on the world scene and the elements of transmission and distribution economics. The last two chapters review the interrelated subjects of rural electrification and the applications of power-system economics to developing countries.

Chapter 1 looks at world energy resources only in so far as is required for the purposes of this book, i.e. to set a scene. It has four underlying concepts. Firstly, knowledge of energy resources is like an inverted pyramid; as we move upwards from the origin-of-fuels to the utilisation-of-energy our knowledge widens not diminishes; we must view data on fuel resources with great humility. Secondly, every aspect of the energy scene is dynamic and all theoretical concepts and data analysis must allow for continuing change. Thirdly, any evaluation of fuel resources rests on current

geological knowledge, engineering capability and economic
feasibility; i.e. changing the demand patterns for fuels significantly alters the value of different fuels and therefore the amounts
worth developing. Fourthly, the perceived value of any resource
varies with the observer; this factor may transcend all economic and
operational perceptions of what is optimal.

Chapters 2 and 3 show how the shortage of capital and the scarcity
and high cost of fuels require the use of economic principles to
produce and consume electricity efficiently, while conserving these
basic resources. Long-run marginal costing should be the basis of
power-sector tariff levels and structures, provided investment rules
are sound. There are interrelationships between optimal reliability
of electricity supply, optimal pricing and optimal investment. An
inter-disciplinary approach is necessary, mainly between economists
and engineers.

Electricity demand forecasting, a subject linked to electricity
pricing (Chapter 9), is dealt with in Chapter 4. Electricity demands
are closely linked to social, economic, technical and industrial
trends. Europe has built up a substantial experience in tackling
forecasting problems, using both large-scale econometric models to
give an 'aggregate' forecast on the one hand, and extensive, detailed
load and market research models to provide a disaggregated forecast
on the other.

Chapter 5 shows how national energy strategies affect the choice
of generating plant-mix by type of fuel. Many governments are
responsible for a co-ordinated and efficient energy sector which can
provide an adequate and reliable supply of energy at minimum cost
in national resource terms, i.e. consistent with overall national
economic strategy. Uncertainty and extremely long lead times for
any major energy sector investment such as power stations or new coal
mines necessitate keeping options open for as long as possible and
creating new options likely to be more cost effective. The future
mix of coal, oil, gas, hydro, nuclear and 'new' energy sources like
solar, wind, wave, biomass etc., is likely to be determined by a
combination of these national economic factors and social preference.

The principles of energy-sector and power-sector modelling are
covered in Chapters 6 and 7 respectively. Economic planning takes
place at two levels. The first level is from the viewpoint of the
national economy and the energy sector. From this viewpoint the power
sector is important, but many other sectors also have a rôle to play.
The second level is from the viewpoint of the power sector itself.
From this viewpoint power-sector planning can be decomposed into
ranges of subproblems, which is fortunate for the modeller. At the
first level, a dynamic simulation model for long-term energy demand
projections must take into account the interaction of supply and
demand throughout the market mechanisms of the whole economy. It
must include such input variables as gross domestic product, energy
prices, fuel taxes, subsidies for conservation measures, measures
concerned with life styles and market shares of alternative energy
forms. At the second level the range of subproblems includes the
planning of generation, transmission and distribution expansion.

Decomposition allows many analytical techniques and refinements to be used; the technique of linear programming seems likely to remain the most promising.

Chapter 8 deals with methods of justifying generation projects. The method to use in a particular case depends on the social, regulatory, statutory and commercial environment, which will normally be different in developed countries and developing countries. The history of the European power sector has profoundly influenced the basis on which new generation proposals are justified today. From the economic standpoint it will almost certainly be necessary to install new hydro or nuclear plants in the immediate future because of the increasing cost of fossil fuels in real terms. This is true for most other systems.

Chapter 9 describes how electricity pricing policy is derived from a mixture of statutes, economic and financial requirements of the power utility and of the government, sound commercial practice and consumer expectations. Load management to suit the economics of both power utility and national economy can be achieved by encouraging consumers to modulate their demand so as to minimise their electricity bills. Incentives to consume electricity off-peak rather than on-peak have an important impact both on consumers and the power sector.

Chapter 10 introduces the concept of geography by dealing with the complex subject of power networks, whose structures have changed appreciably throughout history. Networks are shaped by economic needs, but with three overriding functions: (i) to transfer energy to load centres; (ii) to enable one generating station to stand by for others; and (iii) to interconnect generating stations and loads so that only the stations with the lowest running costs are operated at any time. Increasing distances and magnitudes of energy transfers have been matched by corresponding increases in network voltage level, resulting in 400kV/500kV systems and 110kV/130kV systems being fairly general. Recently 750kV/1100kV has been introduced to interconnect large concentrations of load and generation, using either alternating current (AC) or direct current (DC). Factors increasingly influencing network economics are environment, conservation of energy and reliability. As with generation, all network economics requires appraisal of the numerous possible alternative developments to determine a least-cost solution, having first established criteria of component design and performance. Factors affecting the selection of alternative developments include choice of technology, voltage, conductors, routing and insulation levels. The evolution of distribution networks normally results in changes over time from simple radial feeders to radial distributors, and finally to ringed and/or meshed networks. Power can be 'injected' from the higher to the lower voltage network as a normal means of reinforcing any part of the lower voltage network, wherever and whenever load growth creates a shortage of capacity. However, a replacement problem arises as a distribution network ages and therefore becomes increasingly unreliable, and at some point the cost of more frequent maintenance becomes excessive and there must be

widespread network replacement. This replacement of very old
distribution networks is becoming a very real problem, exacerbated
by the difficulty of obtaining finance solely for the replacement
and rehabilitation of the network, i.e. rather than for load growth.

Chapter 11 deals with rural electrification, especially with
respect to developing countries. It is always an important part of
any rural or regional development plan, and its characteristics are
different from those of urban or industrial electrification: low
density loading; low equipment utilisation; low growth rate of both
the number of consumers and the electricity consumption per consumer;
low load factors; high capital costs per kW installed; high operating,
maintenance and fuel costs per kWh consumed; low revenue from
consumers; and the fact that most (85%) of the 2.5 billion people
in the world who live in rural areas[2] are in developing countries.
These characteristics have serious consequences for both power
utilities and governments, since there will be poor financial and
economic returns from rural electrification schemes especially in
the early stages of development. Pricing policies will also be
difficult because rural consumers will normally be very poor, again
especially in the early stages of development. Subsidies will be
needed to keep rural electrification financially and economically
viable. The economics of rural electrification call for some
cost/benefit analyses to decide upon which consumers, if any, should
be connected to the power system and the timing/staging of such
connections.

Chapter 12 deals with the application of power-system economics
to developing countries. This merits special checklists plus a
handbook for power-system planners who have not dealt with developing
countries. Normally, to obtain aid from governments and lending
agencies a power-sector development programme must be prepared. The
suggested handbooks would itemise such factors as formats for
feasibility and project preparation reports; standardisation of
engineering; financial and economic methods; and the methodologies
for project justification. Developing countries are economically
very sensitive to the recent drastic increases in primary fuels costs,
especially those countries which have no substantial primary fuels
of their own. Since the power sector is one of the most important
sectors in any developing country, the subject of power-system
economics may be even more important for these countries than for
developed countries.

REFERENCES

[1]   MUNASINGHE, M.: 'Power system economics, pricing and
reliability standards'. Paper at the International Symposium
on Electricity Economics and Load Management, Imperial College
London. Reproduced by permission of Imperial College, London,
Power Systems Group

[2]   World Bank Sources, e.g. Annual Reports, World Bank
Atlases, etc.

Chapter 1

Energy resources, sources and uses

1.1 THE ENERGY BACKGROUND

1.1.1 Introduction

Possibly more than any other sector, the power sector is directly susceptible to the world energy scene, in that it is both a user of fuels and a competitor with other forms of energy. In this chapter we look at the world energy scene within which power sectors must operate. Any actual numbers are only important here in so far as they illustrate major constraints on the power system. For exhaustive examination of the energy sector the reader is referred to the published literature.[1]

Fuel availability and price are two major constraints on power systems today. The pre-1973 era of cheap and abundant fuel supplies, mainly oil, is over; possibly for ever. Before 1973 all but the largest, best situated hydro stations were ruled out because of high capital cost; no attention was focused on other renewable energy forms, e.g. wind, wave, solar, biomass etc. Diesel engines/ generators became popular, especially in developing countries where capital was scarce and only small inexpensive machines running on (then) cheap oil were needed. Thus many diesels were purchased and are now causing serious problems with respect to obtaining spares, providing maintenance facilities and getting the foreign exchange for their fuel.

The 1973 price rise was followed by a further price rise in 1979/80 as shown in Fig. 1.1. The price of petroleum is now about fifteen times what it was in 1972 in money terms; five times what it was in real terms. Even 10 years ago such a situation would have been unthinkable.

The important factors which determine the 'resource cost' of fuels today are:

(i)   possible fuel shortages
(ii)   possible even higher fuel prices
(iii)   large capital investments needed to develop new energy
        sources.

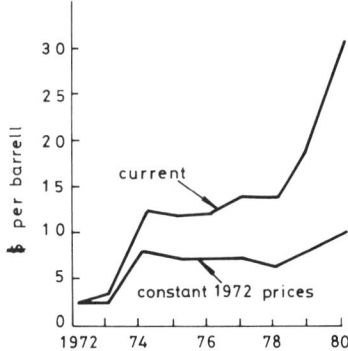

Fig. 1.1   Petroleum prices 1972-80
a: OPEC government sales prices weighted by OPEC output.
b: Deflated by manufactured export prices.

UN Development Forum Business Edition, 31 Mar. 1981.
Reproduced by permission of the editor.

There is still talk of an energy 'gap' in the 1990's between the energy supplies available and the energy demand needed. Periodically such talk becomes important in forcing up fuel prices. New energy sources in the form of renewable fuels are in the news; hydro power has been back in investment programmes for many years; wind, wave, solar and biomass schemes are being developed in many countries. Nuclear development is only being held back for political, social and environmental reasons; experiments with nuclear fusion continue. The cost of utilising most renewable energy is considerable and can only be borne by the developed world.   Though more than half the renewables are in developing countries they have not the capital or the skill to develop them.

Developing countries today are tackling another class of fuel, i.e. 'non-commercial' or 'traditional' fuels, including animal and human labour, animal and vegetable wastes, and fuel-wood.[2] These provide 90% of energy in some countries and 50% in many.   Their growing use has environmental consequences only now being costed; e.g. 60% of countries are not fully replacing the biomass fuels of animal and vegetable wastes.[3] Most non-commercial fuels used are also presently extremely inefficient, having heat-conversion efficiencies of only 1% to 3%.

A possible fuel scarcity and even higher fuel prices are the main causes of the present drive for energy conservation and energy management. A third important factor is shortage of foreign exchange for new energy production equipment and for imported fossil fuels.

1.1.2 Non-renewable fuels

The world economy is presently firmly based upon non-renewable
fuels of coal, oil, gas and uranium. We note:

First, our knowledge of new-renewable resources resembles an
inverted pyramid; the further we move 'up' from the geographical place
of origin of the fuel towards its utilisation, the more we know about
the fuel. Data at the apex, i.e. resource data, must always be treated
with scepticism.

Secondly, energy situations are dynamic. Resource data have
validity only when the calculation is being made since they depend
on: geological and geophysical knowledge about the fuel at a
particular time and place; engineering capability of extracting that
fuel at that particular time and place; and the economic sense of
developing and transporting that fuel from that particular place at
that time.

Thirdly, we must constantly check our forecasts for fuel supply
against those for fuel demand and vice versa; it is easy to allow
these forecasts to 'run away' from common sense when taken separately.
Bringing the two together is the way of ensuring that we make a
reasonable range of forecasts.

Fourthly, fuel resources have different values from different
viewpoints, e.g. of governments, multi-nationals, power utilities
and consumers. Our viewpoints here are that of government/economy
and, to a lesser extent, the power utility.

Allowing for the above, we take a view about non-renewable fuels
likely to be usable now and in the immediate and the far future.
However, there are no universal definitions of fuel resources; those
given below are the easiest to understand of those in current use:[4]

(i)   Resource base is the amount of fuel occurring in a genuine and
       recognisable form.

(ii)   Resources are the amounts of fuel recoverable for the benefit
       of the world.

(iii) Reserves are the amounts of fuel recoverable in terms of (i)
       and (ii) above and:
       (a) Possible reserves are estimated amounts recoverable when
           the knowledge about the fuels is imprecise.
       (b) Probable reserves are estimated amounts recoverable when
           much information is known about the fuels.

(c) <u>Proved reserves</u> are estimated amounts recoverable under present-day economic, institutional and operational conditions. These can be determined fairly precisely for some fuels.

(d) <u>Additional resources</u> make up the gap between (iii)(a) and (i).

(iv) <u>Cumulative production</u>; this is self explanatory and refers to a particular time and geographical location.

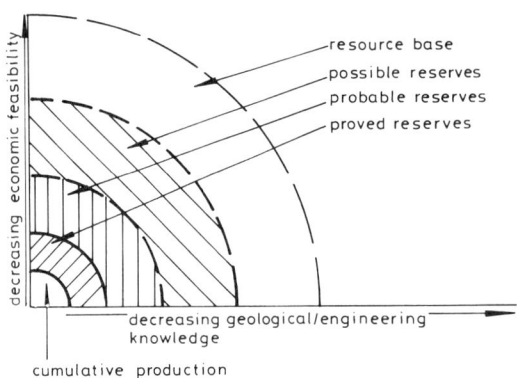

Fig. 1.2   Basic concepts of resource evaluation

ION, D.C.: 'World energy resources, sources and uses'. Paper given at the International Symposium on Electricity Economics and Load Management, Imperial College, London, 24-28 Mar. 1980. Reproduced by permission of Imperial College, Power Systems Group.

In Fig. 1.2 the boundaries are constantly changing. By definition we are continually moving towards Resource Base. Again, Cumulative Production changes the Proved Reserves; technology converts Probable Reserves into Proved Reserves; research expands Possible Reserves from Additional Resources. Economic feasibility is changing and higher fuel prices convert previously uneconomic into economic development, subject always to geology and technology; a higher fuel price per se never <u>automatically</u> increases reserves.

Geology and technology both change with knowledge, sometimes quite suddenly, e.g. following the finding of new large gas/oil fields. Areas considered worthwhile prospecting are changing. Fig. 1.3 gives the world areas considered worthwhile for oil prospecting; in 1947 oil prospecting was confined to land, by mid-1950's off-shore working began, and by 1970 off-shore deep-water working was established.

The 'preferred' quality of fuels varies with time/place. Coals vary from soft peats to hard anthracites; at some time/place each type

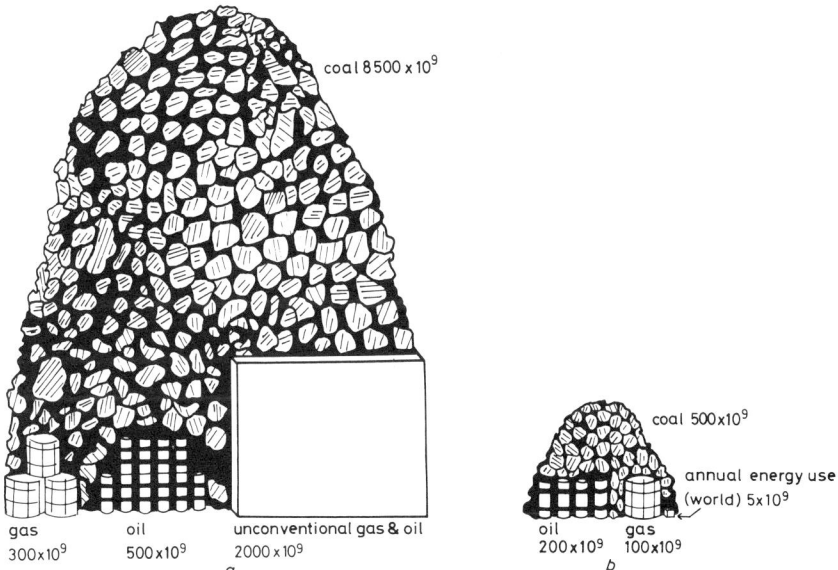

coal 8500 x 10⁹

coal 500x10⁹

annual energy use (world) 5x10⁹

gas 300x10⁹	oil 500x10⁹	unconventional gas & oil 2000 x 10⁹	oil 200x10⁹	gas 100x10⁹

a   *b*

Fig. 1.4   Estimated and proven resources of fossil fuels

> a: Estimated resources total 11300×10⁹ (or 11300 billion) tonnes of coal equivalent
> b: Proven resources total 800×10⁹ (or 800 billion) tonnes of coal equivalent
>
> Coal is not a new or renewable source of energy but it must not be forgotten that it provides the world's backstop if all the alternative energy sources fail – or, more likely, take a lot longer to fix than anyone suspects. The reserves of coal are huge in absolute terms and dwarf those of oil, gas and the unconventional hydrocarbons – oil shales and tar sands. The graph above, based on Nature 278, D.O. Hall, depicts the scene under 'estimated reserves'. As food for thought it should be pointed out that photosynthesis produces 8 × 10¹⁹ tons of usable carbon in biomass <u>every</u> year, ten times world energy consumption.
>
> UN Development Forum Business Edition, 17 Apr. 1981. Reproduced by permission of the editor

is 'preferred'. The scarcity of good coking and industrial coals has led to the use of blends in industry considered impossible twenty years ago. Again, a barrel of heavy high-sulphur crude has the same volume as one of light low-sulphur crude; it has a different weight and, more importantly, a different economic value. The economic value of light crude is high in the USA because its high gasoline yield

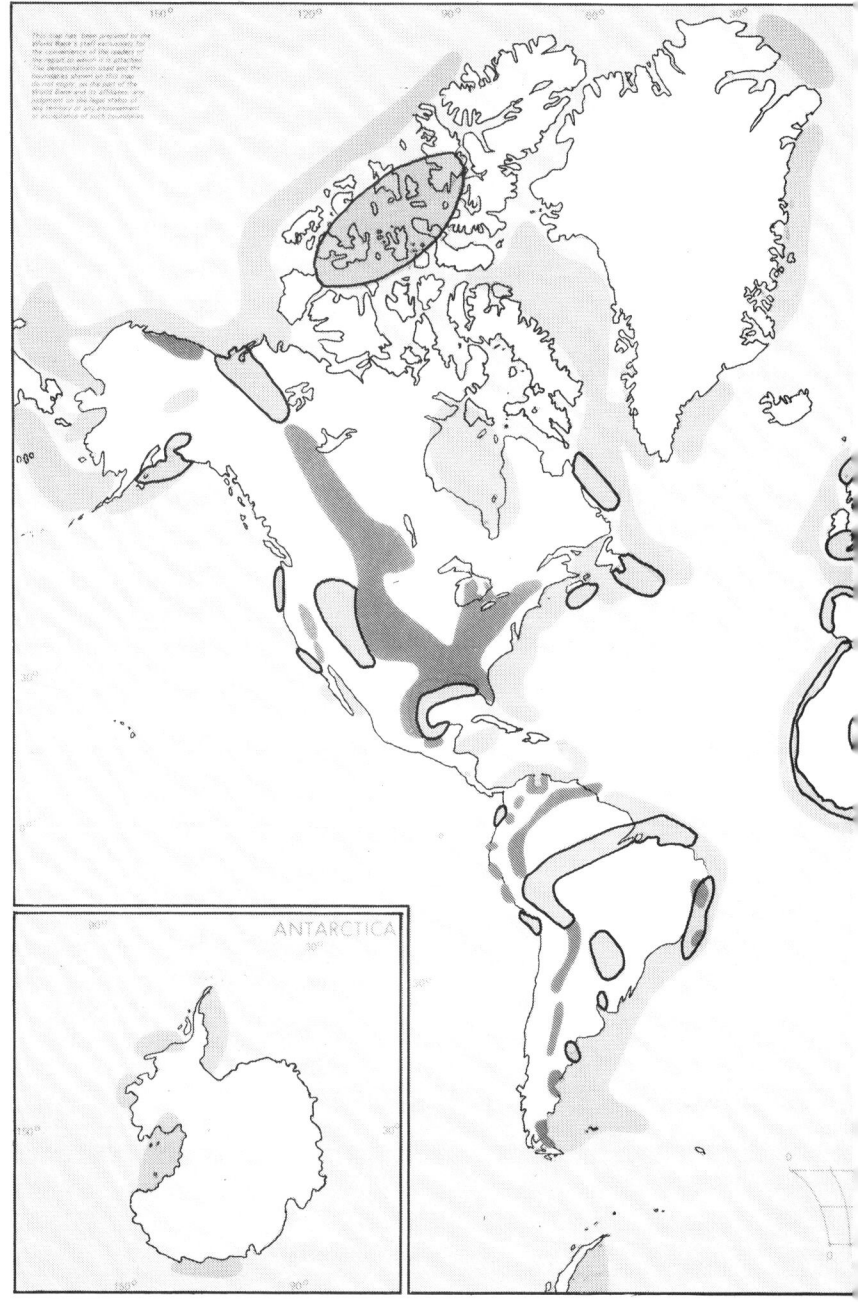

Fig. 1.3   Petroleum basins of the world

'A program to accelerate petroleum production in the developing countries', World Bank, Jan. 1979. Reproduced by permission of the World Bank.

PETROLEUM BASINS OF THE WORLD

PRODUCTIVE

PARTIALLY EXPLORED

UNEXPLORED

matches the high gasoline demand there. Light crude has less value
in Europe where less gasoline and more fuel-oils are used, but it
has a high value in Japan because of low sulphur content, Japan having
strong anti-pollution constraints. Low-sulphur crude has little
value in sparsely populated central Siberia.

Fig. 1.4 shows the estimated world proven non-renewable fuel
resources alongside present annual usage. Coal predominates with
large Additional Reserves which, although not yet firmed up, makes
coal the world's fuel 'back-stop'. Proved Reserves of shale-oil and
tar-sands are unlikely to play a significant role until the 1990's.

1.1.3 Proved reserves of non-renewables

(a) Coal

The International Energy Agency is presently setting up a world
coal data bank which is badly needed. Fig. 1.5 shows one recent view
of world coal resources. The term 'Recoverable Reserves' is a
coal-industry definition of looser quality than the Proved Reserves
defined earlier. Recoverable Reserves means reserves considered
'economically recoverable' in which the USA, China and the USSR lead.

Fig. 1.5 can be used for indicating the rôle to be played by coal,
even though the data from which it is derived are not of equal
validity. Most estimates of China's coal resources are based on very
old records; the USSR's data are not precise; the USA data are
challenged because of the high 'rate of coal recovery' assumed.
Again, in 1974 the UK National Coal Board reported its Proved Reserves
to the World Energy Conference as 4 gigatonnes (1 gigatonne(Gt) =
10^9 tonnes). In 1976 it reported the 4 Gt but also added the figure
of 45 Gt as an alternative. Recently only the 45 Gt figure is reported,
as shown in Fig. 1.5. 45 Gt is more compatible with Proved Reserves
in other countries, the 4 Gt being 'Operating Reserves' recoverable
from existing mines. We could compare Operating Reserves of coal
with the Working Stocks of crude oil reported from the USA. The 45
Gt figure at the present coal usage of about 0.15 Gt per year would
mean coal in the UK would last for about 300 years. However, the
4 Gt figure as Working Stock with production at 0.13 Gt a year gives
assured coal for only 30 years. The latter interpretation gives a
lead time of at least 10 years to get new mines into operation. The
UK case is presented to show that coal data handling needs a lot of
thought and common sense.

The USA, China and the USSR are increasingly requiring their coal
for their own use. Australia, Canada and South Africa have small
local demands and are potential major coal exporters; they have large
Proved Reserves but their supplies are a long way from likely markets.
Therefore, it is doubtful whether there will be a global trade in
coal similar to that in oil.[5] One key to such a coal trade lies with
the oil multi-nationals who already play a large rôle in the world
coal scene.

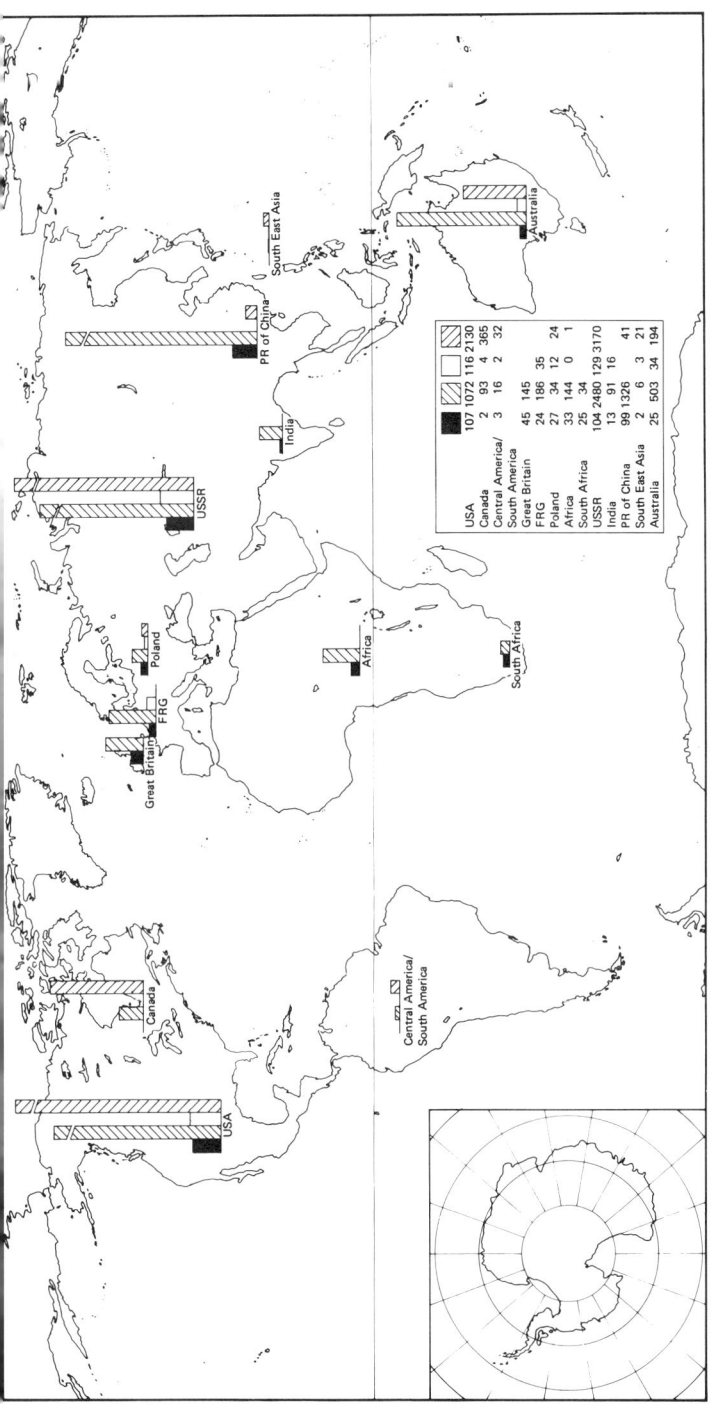

	107	1072	116	2130
USA	107	1072	116	2130
Canada	2	93	4	365
Central America/ South America	3	16	2	32
Great Britain	45	145		
FRG	24	186	35	
Poland	27	34	12	24
Africa	33	144	0	1
South Africa	25	34		
USSR	104	2480	129	3170
India	13	91	16	
PR of China	99	1326		41
South East Asia	2	6	3	21
Australia	25	503	34	194

Fig. 1.5　Coal reserves and resources

Recoverable reserves (10^9 t)　　　Bituminous coal/anthracite

Additional resources (10^9 t)

Recoverable reserves (10^9 t)　　　Subbituminous coal/lignite

Additional resources (10^9 t)

'Survey of energy resources 1980. Enclosure 1.2,
11th World Energy Conference, Munich, 8–12 Sept. 1980.
Reproduced by permission of the World Energy Conference.
© World Energy Conference.

(b)   Petroleum

Accurate resource evaluation for oil is even more difficult than for coal; oil is mobile, occurs in many more places, is less easily discoverable, and the amount recoverable can vary over a range 5% to 60% of the total. Proved Reserves of oil are more comparable with Working Stocks than the Proved Reserves of coal. Giant oil/gas fields play a more important rôle than giant coal fields. Some estimates place about 35% of total world recoverable oil in some 15 giant fields.

Fig. 1.6 shows the world's resources of crude oil. The Middle East has the lion's share: 55% of the total.   The Organisation of Oil Exporting Countries (OPEC) has some 70% of Proved Reserves and about 60% of estimated world ultimate Recoverable Reserves.   The USSR may be running into difficulties; one single field at Samatlor holds probably about 70% of Proved Reserves.   The USSR has doubled oil production over the last 10 years to become the world's largest single producer; however, Fig. 1.6 shows the USSR with a large Resource Base, and sufficient Proved Reserves are likely to be firmed up to support the present production level of 0.6Gt a year.   The amount of Proved Reserves of oil the USSR has is important for the rest of the world. A small Proved Reserves would mean the USSR needs to get some oil from elsewhere; thus less oil would be available for the rest of the world and there would be strategic implications.

Fig. 1.7 shows the world's resources of natural gas. Natural gas is different in many ways from oil. Types of gas are:

(i)   Gas associated with crude oil in an oil field
(ii) Gas not associated with crude oil but found separately, to
      include:
      (a)   gas with a common origin to oil
      (b)   gas formed naturally from coal.

In extreme cases as much as nine-tenths of natural gas can be unburnable because of its chemistry. Transport of gas is costly, requiring pipelines or special tankers and terminals. Thus the financing and structure of the natural gas industry is different from the oil industry, because long-term contracts are necessary to justify the massive investment in fixed assets.

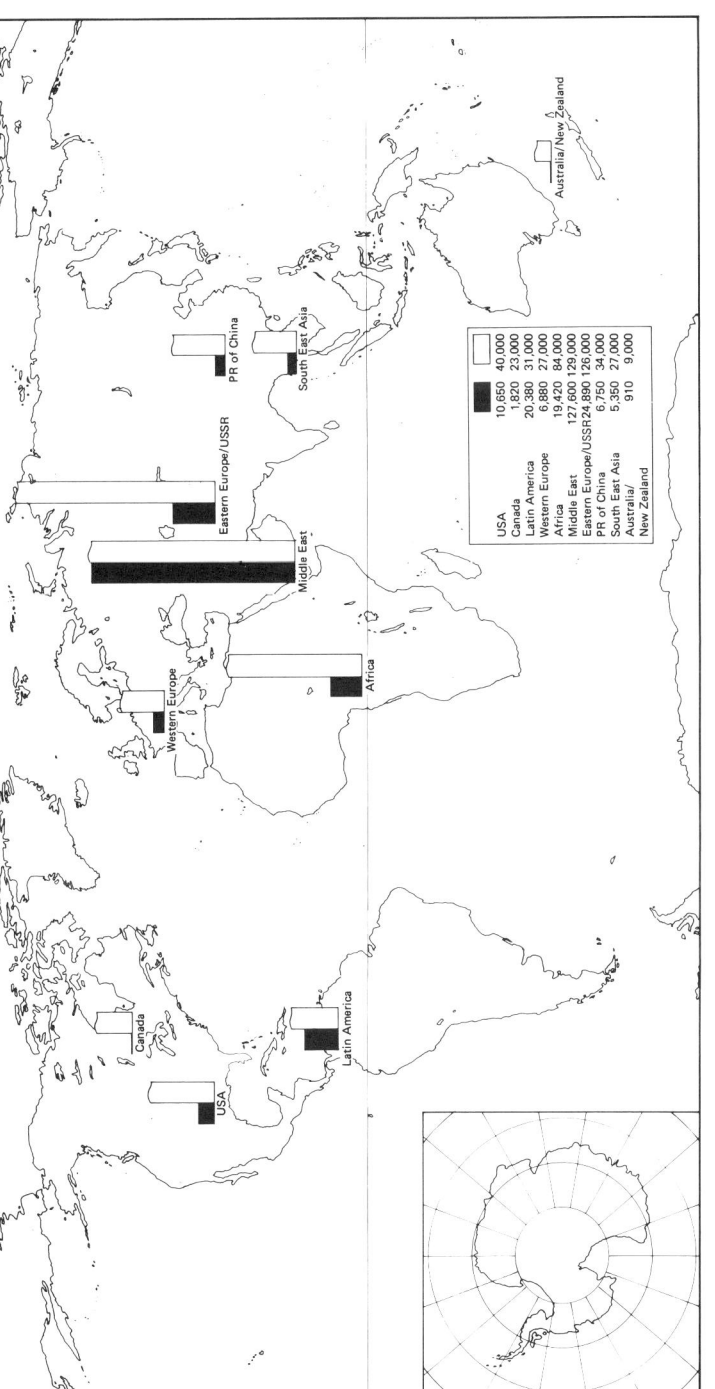

Fig. 1.6 Crude oil and natural gas liquids reserves and resources

☐ Proved recoverable reserves in place (10^6t)

■ Estimated additional resources in place (10^6t)

Country/Region

	☐	■
USA	40,000	10,650
Canada	23,000	1,820
Latin America	31,000	20,380
Western Europe	27,000	6,880
Africa	84,000	19,420
Middle East	129,000	127,600
Eastern Europe/USSR	126,000	24,890
PR of China	34,000	6,750
South East Asia	27,000	5,350
Australia/ New Zealand	9,000	910

40% recovery factor supposed

Survey of energy resources 1980, Enclosure 2.2, 11th World Energy Conference, Munich 8–12 Sept. 1980. Reproduced by permission of the World Energy Conference, London. © World Energy Conference.

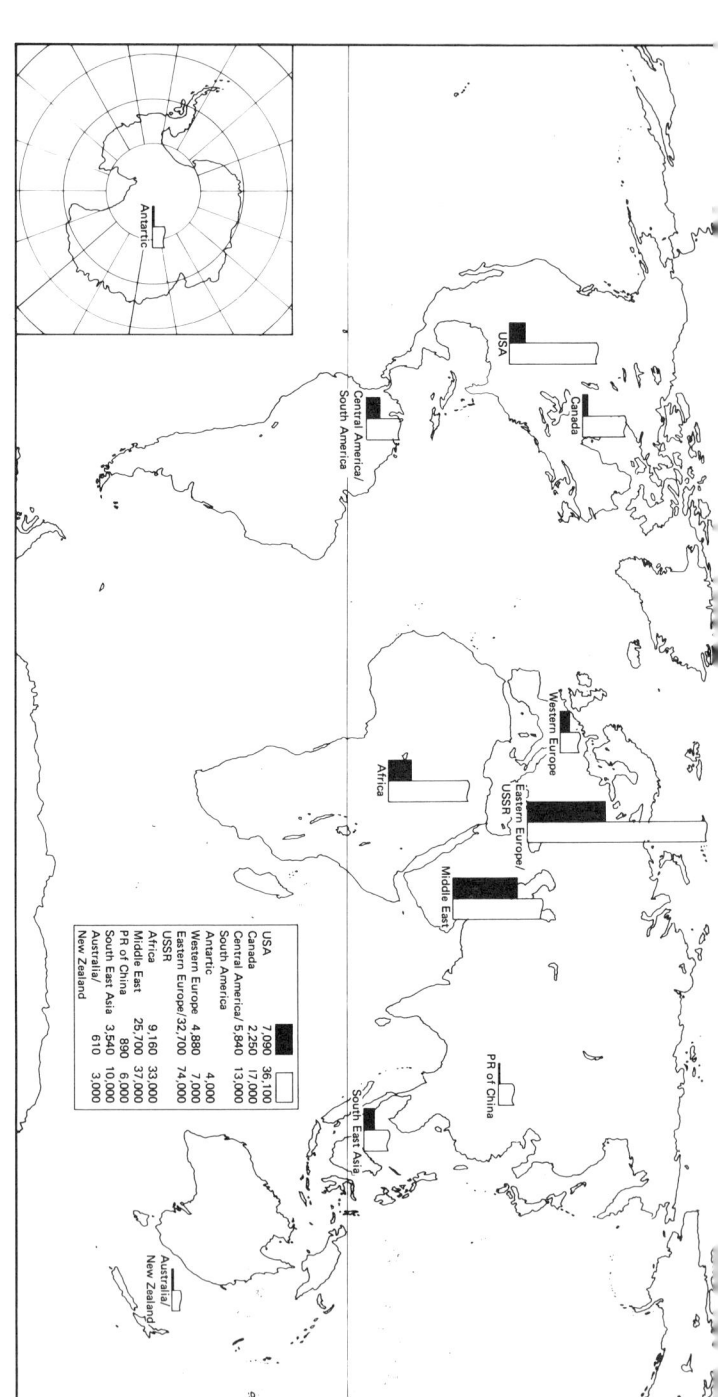

Fig. 1.7 Natural gas, proved reserves and estimated additional resources

Proved recoverable reserves in place ($10^9 m^3$)

Estimated additional resources in place ($10^9 m^3$)

Country/Region

80% recovery factor supposed

Survey of energy resources 1980, Enclosure 2.7, 11th World Energy Conference, Munich 8-12 Sept. 1980. Reproduced by permission of the World Energy Conference, London.
© World Energy Conference.

	Proved	Estimated
USA	7,090	36,100
Canada	2,250	17,000
Central America/ South America	5,840	13,000
Antartic		4,000
Western Europe	4,880	7,000
Eastern Europe/ USSR	32,700	74,000
Africa	9,160	33,000
Middle East	26,700	37,000
PR of China	890	6,000
South East Asia	3,540	10,000
Australia/ New Zealand	610	3,000

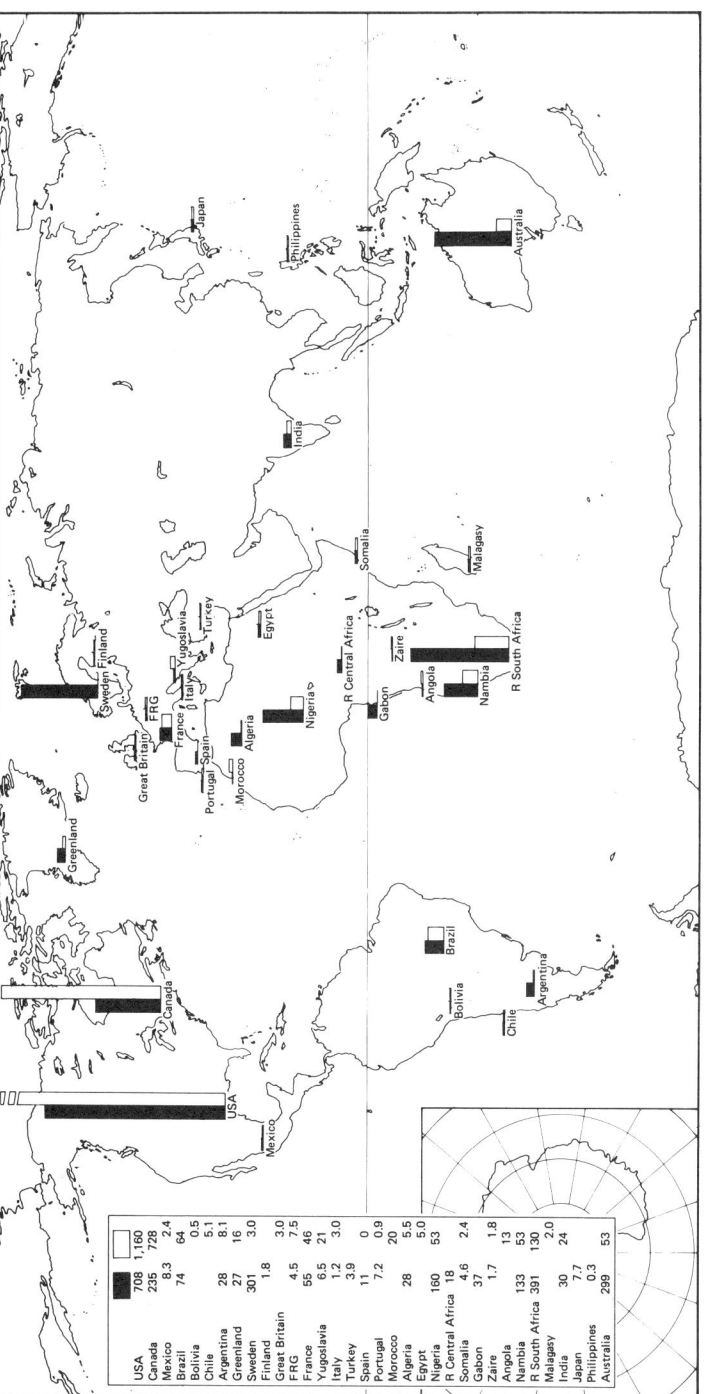

Fig. 1.8 Nuclear resources, uranium reserves and resources

Reserves = reasonably assured resources ≤ 130 $/kg U

Resources = estimated additional resources ≤ 130 $/kg U

Country/Region

Survey of energy resources 1980, Enclosure 3.3, 11th World
Energy Conference, Munich 8-12 Sept. 1980. Reproduced by
permission of the World Energy Conference, London.
© World Energy Conference.

	■	□
USA	708	1,160
Canada	235	728
Mexico	8.3	2.4
Brazil	74	64
Bolivia		0.5
Chile		5.1
Argentina	28	8.1
Greenland	27	16
Sweden	301	3.0
Finland	1.8	
Great Britain		3.0
FRG	4.5	7.5
France	55	46
Yugoslavia	6.5	21
Italy	1.2	3.0
Turkey	3.9	
Spain	11	0
Portugal	7.2	0.9
Morocco		20
Algeria	28	5.5
Egypt		5.0
Nigeria	160	53
R Central Africa	18	
Somalia	4.6	2.4
Gabon	37	
Zaire	1.7	1.8
Angola		13
Namibia	133	53
R South Africa	391	130
Malagasy	30	2.0
India	7.7	24
Japan	0.3	
Philippines		53
Australia	299	53

(c) Fissile fuels

Fig. 1.8 shows world uranium resources. Exploration for thorium awaits the 'fast' reactor. Fig. 1.8 shows that the world's uranium industry has a unique 'Reserves' and 'Resources' nomenclature based on forward costs of extraction of uranium oxide, measured in kilogramme weight, taken after exploration but with no allowance for taxes and profits. The reasonably Assured Resources of Fig. 1.8, taken at under US$130 per kg weight of uranium oxide, roughly equate with the Proved Reserves of other non-renewable fuels. Exploration for uranium ore has been spasmodic because demand has fluctuated widely; and few countries are involved and usually in response to their own direct needs.

1.1.4 Renewable but exhaustible fuels

Fuel-wood is the world's most important renewable fuel which may be exhaustible. It is the third largest single energy source after oil and coal and is still first for many consumers; in some areas it is the only energy source. In many developing countries fuel-wood accounts for over half of total energy sources.

Fuel-wood has rarely, if ever, been used to generate electricity, and in the future fuel-wood power stations are possible but not probable for technical reasons, except under special circumstances.

In the future the greatest use for renewable but exhaustible fuels may lie in gas-oil for motor cars by extracting alcohol from wood and mixing it with gasoline. In 1980 some 18% of Brazilian cars were using a 20% alcohol/80% gasoline mix and the use of this mix is expected to increase. Nonetheless, it is difficult to see how gas-oil could be a major automobile fuel.

The other important renewable but exhaustible fuel is biomass, which uses animal and vegetable wastes to generate gas for a variety of purposes, including generating electricity in small quantities but mainly in remote rural areas.

1.1.5 Inexhaustible fuels

All other renewable forms of energy can be considered in-exhaustible. To understand properly the part these fuels may play in the world energy scene we note that inexhaustible fuels have the following characteristics:

Firstly, inexhaustible energy sources are diffuse and must be concentrated to be converted into useful energy. A good example is solar energy. A small energy input is needed for steering the apparatus used for the concentration process.

Secondly, the 'Resource Base' for all inexhaustibles is, by definition, infinite. This means only specific local data has any meaning, i.e. we cannot talk about 'world' amounts.

Thirdly, exploitation of inexhaustibles involves environmental costs. Large areas of solar collectors on land, at sea or in space, hundreds of wave generators in off-shore chains, or massive wind generators on hilltops may be as great an environmental hazard as one large thermal power station or hydro reservoir, and both would supply more energy than the inexhaustibles.

Fourthly, small units/single schemes may have an earlier/greater energy impact than massive/multiple schemes. Massive hydro-electric plus irrigation schemes, although very fashionable, rarely benefit the local population except a few linked directly with the irrigation. Massive solar or wind schemes do not help the local people; all the energy disappears into the electricity network mainly for industry and the towns. On the other hand, small inexhaustible units, locally manufactured with minimum imported materials, simply designed and easily maintained would have an immediate effect on the local people. Such schemes, e.g. micro-hydro generators, could enable a rural economy to rapidly 'take-off' economically.

Fifthly, all inexhaustible forms of energy are better known in the 1980's than in the 1970's; practice has almost caught up with theory.[6] The inexhaustibles are often at the threshold of technological development and have had the same impact on leading scientists of the 1970's as had nuclear physics in the 1950's. However, coal, oil and natural gas still hold considerable challenges. Putting bright people to work on these fuels seems of more immediate value than putting them on to inexhaustible forms of energy; this balance may change around the year 2000.

(a) Tidal power

Only about 30 universally recognised sites exist in the world where a natural concentration of tides seems suitable, with modern technology, for converting tidal power into economic energy. There are many sites for other technologies, e.g. small tidal mills, but tidal power is never likely to be a major source of world energy.

(b) Wave power

The problems with wave power are now better appreciated than in the enthusiastic 1960's, but costs must be reduced drastically to make a significant economic impact from this source of energy anywhere in the world; wave power is unlikely to be a major world energy source this century.

(c) Geothermal power

Most exploitations of geothermal power use 'heat seepages'. Although at least one 'buried heat field' in Italy has been found,

it has not yet been used commercially. It is only by drilling that
commercial fields can be found and heat reserves proven. There seems
little doubt that in future years low-grade heat sources and smaller
temperature differences in the steam/water will be increasingly
utilised. However, whether schemes are large or small, on land or
in the sea, the environmental impact will require careful monitoring.
Some geothermal sources may prove to be exhaustible rather than
inexhaustible; other schemes upset delicate natural balances in the
earth's crust and eventually have to cease production.

(d) Hydro power

Approaching one quarter of the world's electricity is generated
by hydro plant, but the potential for big new economic schemes is
limited; many are far from load centres; all require large capital
investment. Micro-hydro units are likely to have an immediate
potential in remote rural areas. The pumped-storage type of hydro
station is often an economic means of extending the base load and
shaving the peak.

1.1.6 Special aspects for developing countries[7]

Estimates of energy and fuel resources in developing countries made
prior to 1973 may no longer be relevant. This applies to all energy
resources, but especially to coal, natural gas and hydro. Many
developing countries resemble China, where currently used estimates
of coal reserves were made long ago; and most estimates for hydro
resources pre-date 1973. Hydro potential includes two different
concepts which can cause confusion for planning purposes. The first
concept is 'potential' power, i.e. maximum hydro power which could
be generated from all hydro resources of a country in a normal
hydrological year; the potential powers at individual hydro sites
are simply added. Of more practical use is the 'economic' power,
i.e. the proportion of 'potential' power economically worth
developing at a given state of technology and a given cost of
alternative energy sources. It is important to consider what the
alternative energy source would be, the normal being thermal power
from oil prior to 1973. Cheap oil automatically ruled out most
potential hydro sites and the inventory of potential hydro projects
was short. Today these inventories are being updated in many
developing countries. However, much exploratory work needs to be
done on site and updating takes time.

Estimates of potential resources of non-renewable fuels are also
out of date in many developing countries. A few years ago the World
Bank reviewed the potential resources of hydrocarbons suitable for
development in some smaller developing countries[8] already known to
have some potential resources, but in the 1950's and 1960's these
were not considered enough to carry out detailed surveys to turn
Potential Reserves into Proved Reserves. Production costs from such

reserves were unlikely to compete with Middle East oil then at US$1 to US$2 per barrel. The study produced an interesting list of countries which had potential hydrocarbon resources; although small by world standards and without the giant oil/gas fields mentioned earlier, for an individual country's needs these resources could make a substantial contribution to the economy by cutting down imported oil and corresponding balance-of-payments deficits.

As we have seen, in many developing countries a large proportion of the energy needs are met from non-commercial sources, e.g. animal and human labour, fuel-wood, vegetable and animal wastes. In these countries fuel-wood remains the most important energy source; for the poorest people it represents the only energy source. It is difficult to measure the potential resources of non-commercial energy and any estimates are of doubtful accuracy.

Commercial forms which substitute for non-commercial energy are kerosene, natural gas, and sometimes electricity, possibly at heavy resource cost to the economy. Since 1973 the increased cost of petroleum and natural gas has tended to remove any natural price incentive to use oil rather than non-commercial fuels. To the people gathering them non-commercial fuels are a free gift whilst paraffin has to be paid for, and at an increasing price. This viewpoint of the energy consumer can be very different from that of the economy. Fuel-wood is not necessarily inexhaustible, its increased use aggravates the problem of deforestation; people extract fuel-wood in excess of the ability of the forest to regenerate it which can lead to the serious problem of land erosion. Again, with the removal of the forest and consequent soil erosion excessive water run-off plus siltation affects the capacity of many reservoirs at adjacent dam-sites. This siltation has a direct impact on hydro capacity, often a major commercial energy source.

The various agricultural and animal wastes have an important alternative use other than for energy, e.g. as fertilisers. Again, because of the increasing cost of commercial fuels like kerosene, people are using more animal dung for cooking and heating, especially in rural areas, thus depriving the soil of a valuable ingredient.

1.2 THE FUTURE WORLD ENERGY SCENE

1.2.1 Introduction

We cannot estimate the future position of any power sector until we have some grasp of the future world energy scene[9], and especially if sudden changes in this scene are likely. Let us look at some scenarios of the future world energy scene in this context.

It now seems unlikely that the rest of the 20th century will be disturbed by insurmountable problems in world energy. There will be temporary fuel shortages and high fuel prices, but no sustained panic of an energy demand/supply 'gap'. Until recently many experts believed we would be distributing energy shortages in the 1990's.

The fuel surpluses of 1980/81/82 make us think again. If serious shortages did occur the political, social and economic consequences would be grave and have a profound effect upon the power sector. Thus it is important to decide whether shortages are likely or not. Any such disaster is unlikely before 2000 and possibly before 2020: this is our conclusion in this Chapter.

1.2.2 Substitution between dominant fuels

   History shows a progression in the dominant fuel. Some time ago in developed countries, coal substituted for the fuel-wood still in use in developing countries today. During the earlier twentieth century, petroleum and natural gas tended to substitute for coal. Today, although coal is staging a comeback, many regard nuclear, hydro power, and renewable energy sources to be now substituting for petroleum and natural gas. All renewable energy sources can be attributed to solar energy, though in time energy from an artificial sun, i.e. from nuclear fusion, may replace all other forms. It is important to realise that this constant substitution is going on, but it is more important to know when a major fuel shift is likely. Therefore, we must pinpoint some future date before which a major shift in dominant fuel type is unlikely to have taken place. We start by assuming that, because of reasons given in Section 1.1, such an important event is unlikely to take place before at least 2020 and we are faced with known technology at least until 2020 with no energy gap before 2000. Fortunately for power-system planners today, the period 1982 to 2000 is the most important one. Despite power-system equipment lasting for as long as 30 to 50 years before it is replaced, putting a time value on money by using discounted cash-flow techniques makes the first ten to fifteen years of the life of any equipment the crucial ones.
   It seems certain also that the developed world will dominate the energy scene at least until our crucial year of 2020; so conservation in developed countries is vital because most energy is used in developed countries and this is likely to continue for another 40 years. A large conservation programme in developed countries could push back the grey area of uncertainty over energy 'gaps' beyond the year 2000, or even eliminate the area of uncertainty altogether. Sufficient energy conservation is likely to be achieved to avoid any prolonged fuel 'gap' before 2000, but this assumes:

(i)   making more efficient use of all fuels and energy than at present
(ii) substituting the more plentiful resources of commercial energy, and those less costly to the national economy, for the less plentiful and more costly sources of commercial energy
(iii) substituting commercial energy for those resources of 'non-commercial' energy more costly to the economy.

By this process coal, nuclear, geothermal, the cheaper forms of hydro and solar energy, and biomass, would gradually substitute for oil and gas mainly in developed countries, and fuel-wood and dung in developing countries.

1.2.3 World energy supply-demand position

To produce a credible world energy supply-demand balance for the future we proceed as follows:

First, we forecast an 'ultimate' amount for each type of energy resource. Here, 'ultimate' means our vital year of 2020. These estimates:
• are more precise than those given earlier in this Chapter; although partly derived from these, other factors are also involved; and
• enable us to deal in some reasonable way with our 'grey' period of 2000 to 2020.

Secondly, we forecast the rate at which these energy resources can be produced.

Thirdly, we forecast the extent to which conservation measures will take place. We apply conservation both on the 'supply' side, e.g. substituting plentiful economically-cheap for scarce economically-dear resources, and also on the 'demand' side, e.g. making more efficient use of all types of energy.

Extrapolations of past trends on the 'supply' and the 'demand' side must fit reasonably well into the world energy scenario which emerges if we are to accept that scenario itself as being 'reasonable'.[10]

Table 1.1 shows such a 'reasonable' prediction for potential world primary commercial energy production until 2020 as measured in normal quantities. Using a common unit of heat in exajoules (1 exajoule(EJ) = 10^{18} joules) is more useful.

In Table 1.1 we note:

(i)   The rapid potential build-up in coal use in actual exajoules, but not as an increasing percentage of the total world potential energy production.

(ii) The potential slow decline in oil use between 1985 and 2000, followed by a rapid decline to 2020.

(iii) The significant share of unconventional oil and gas, e.g. tar-sands and shale-oil, expected in 2020 but not until that year.

(iv) The potential dominant rôle of nuclear energy by 2020 to just under one third of total potential world energy production. This takes the view that in the future there will be less serious obstacles to the installation of nuclear plants than at present.[11]

(v)   The rather disappointing growth potential in hydro plant, despite the 'great expectations' for hydro development of the late 1970's.

(vi)   The important potential in future years for renewable energy, other than hydro; throughout the later period it accounts for about 10% of all potential energy production.

(vii)   The doubling of potential energy production between about 1985 and 2020, i.e. within a period of 35 years (on average at about 2% per annum growth).

Table 1.1 Potential world primary energy production, exa-joules (1EJ = 10^{18}J)*

Resource	1972	1985	2000	2020
Coal	66	115	170	259
Oil	115	216	195	106
Gas	46	77	143	125
Nuclear	2	23	88	314
Hydraulic	14	24	34	56
Unconventional oil and gas	0	0	4	40
Renewable, solar, geothermal, biomass	26	33	56	100
Total	269	488	690	1000

In order to complete our world-energy supply/demand picture we now consider each major fuel as a potential source of energy.

1.2.4 Coal production

Table 1.2 and Fig. 1.9 show that world coal production could rise to approaching 9 billion tonnes per year by 2020, i.e. three times the 1981 level. Table 1.2 shows a breakdown of coal sources. This introduces the geography needed when planners are considering a particular country. The present 'big three' coal producers, China, USA and USSR, are expected to continue to account for about two-thirds of the total world coal production throughout the period to 2020. Other large coal producers in 2020 are expected to be: India, 5% to 6% of world production; Australia, 4% to 5%; Poland and South Africa, both 3% to 4%. These countries are already important coal producers.

---

* Source: KIELY, J.: 'World energy in the 21st century', Chartered Mechanical Engineer, May 1980, Table 7. Reproduced by permission of the Editor

Table 1.2   Estimated future production by the main coal-
producing countries*

Coal production in 10^6 TCE

Country	1975	1985	2000	2020
Australia	69	150	300	400
Canada	23	35	115	200
Czechoslovakia	80	93	100	110
Germany (Fed. Rep.)	126	129	145	155
GDR	76	80	90	100
Great Britain	129	137	173	200
India	73	135	235	500
Poland	181	258	300	320
South Africa	69	119	233	300
Other countries	224	330	449	561
Subtotal	1049	1466	2140	2846
China (People's Rep.)	349	725	1200	1800
USA	581	842	1340	2400
USSR	614	851	1100	1800
Big three subtotal	1544	2418	3640	6000
Total world population	2593	3884	5780	8846

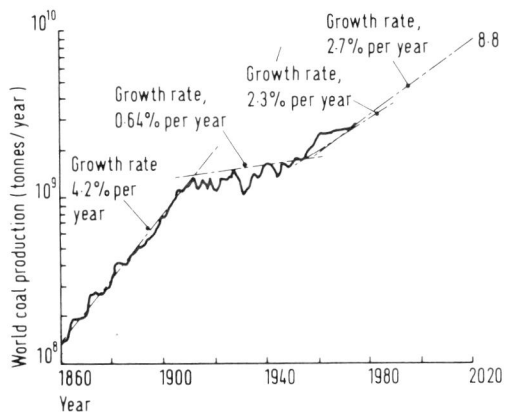

Fig. 1.9   World production of coal (semi-logarithmic scale)

KIELY, J.: 'World energy in the 21st century',
Chartered Mechanical Engineer, May 1980, Fig. 1.
Reproduced by permission of the Editor

---

* ibid., Table 2.   Reproduced by permision of the Editor.

1.2.5 Petroleum production

Table 1.3 gives the potential breakdown of petroleum production
around 2020 consistent with Table 1.1. A preponderance of petroleum
production in the Middle East and North Africa of about 40% of total
world production is still expected, with the USSR, Eastern Europe
and the China 'bloc' next in line with over 20%, and North and Latin
America each about 10%. Instability in the Middle East will thus
continue to be as dominant a factor in the world energy scene in the
future as in the past. As we saw in Table 1.1 it is estimated that
oil production will peak in the 1990's and thereafter decline.

Table 1.3 Estimated ultimate recoverable conventional
           resources of oil*

	Billion barrels	%
North America	200	11
Western Europe	78	4
Japan, Australia, New Zealand	21	1
USSR, Eastern Europe, China	416	24
Middle East, North Africa	764	42
Central and Southern Africa	79	4
East and South Asia	85	5
Latin America	160	9
Total	1803	100

1.2.6 Natural-gas production

Table 1.4 shows the breakdown by region of the potential natural
gas production of the world to the year 2020. It has been assumed
that gas production will not peak until after oil production has
peaked, i.e. in the early part of our 'grey' period of 2000 to 2020.
By 2020 gas production is expected to be about three times what it
was in 1972, about 2.5 times what it was in 1976, and about twice
what it was in 1981. After the peak year of 2000 an increasing
proportion, possibly up to one-half of total production, could be
coming from OPEC, a different situation from today which may give
rise to possible future strategic problems. Up to and including 2000
the largest producer of natural gas will be the Centrally Planned
Economies accounting for about one-third of total production. By
2020 only about half of the world's natural gas resources are expected
to have been used up.

---

* Source: KIELY, J.: 'World energy in the 21st century', <u>Chartered
<u>Mechanical Engineer</u>, May 1980, Table 3. Reproduced by permission of
the Editor.

Table 1.4   Estimated natural gas production capability
            by region (exajoules)*

Region	1976 (actual)	1985	2000	2020
North America	23.0	29.7	27.3	10.7
Western Europe	6.4	9.6	8.7	2.2
Japan, Australia, New Zealand	0.3	0.4	2.1	4.6
USSR, East Europe	12.8	21.8	55.7	28.5
China, Other Asia	1.4	1.7	2.9	6.1
OPEC, Group 1	0.5	7.0	18.1	17.7
OPEC, Group 2	3.4	4.9	21.3	45.6
Central America	0.9	1.1	2.3	1.6
South America	0.8	1.1	2.2	4.8
Middle East	0.1	0.5	1.0	0.3
North Africa	0.2	0.3	0.5	0.5
Africa, South of the Sahara	0.1	0.1	0.2	0.1
East Asia	0.1	0.1	0.2	1.6
South Asia	0.3	0.5	1.0	0.7
World Total	50.3	76.8	143.5	125.0

1.2.7 Nuclear production

It seems likely that nuclear will be a major source of energy by
2020, despite heavy opposition in the period since 1976/77. It also
seems likely that the light-water reactor will predominate, that the
breeder reactor will not be introduced until the late 1980's at the
very earliest[12], and that 'normal' nuclear power installations will
be in reactor units of not less than 600 MW electrical output from
1980 onwards and 1000 MW from 1990 onwards. Table 1.5 shows that
nuclear power could double every 11 or 12 years within our 'grey'
period of 2000 to 2020. More nuclear reactors are presently located
in OECD countries than anywhere else, and this trend is expected to
continue. However, by 2020 as much as one-fifth of the total world
capacity of nuclear plant could be located in the developing
countries.

---

* Source: KIELY, J.: 'World energy in the 21st century', <u>Chartered
Mechanical Engineer</u>, May 1980, Table 4. Reproduced by permission
of the Editor.

Table 1.5   Projected world nuclear power installation (GWe)*

World regional grouping	1975	2000	2020
OECD countries	68	800	2225
Centrally planned economies	7	560	1850
Developing nations	1	180	925
Total	76	1540	5000

Fig. 1.10 shows the effect of the breeder reactor on annual uranium demand; this is significant for planners whose systems are likely to become nuclear orientated.

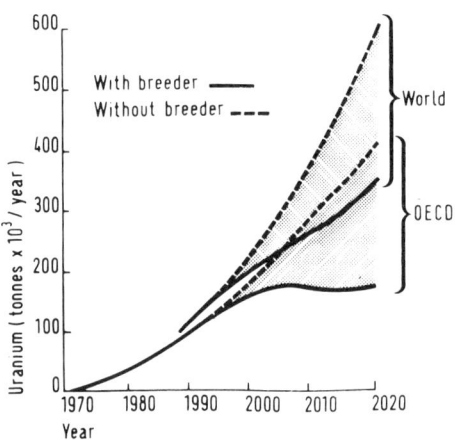

Fig. 1.10   Effect of breeder on annual uranium demand

KIELY, J.: 'World energy in the 21st century', Chartered Mechanical Engineer, May 1980. Fig. 2. Reproduced by permission of the Editor.

1.2.8 Hydro production

Hydro power will be developed as rapidly as is environmentally,

---

* ibid., Table 5. Reproduced by permission of the Editor.

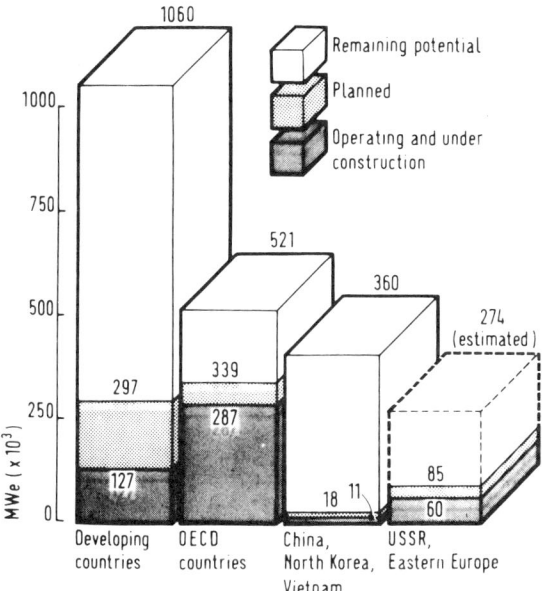

Fig. 1.11   World total installed and installable hydraulic
            capacity

            KIELY, J.: 'World energy in the 21st century',
            Chartered Mechanical Engineer, May 1980, Fig. 3.
            Reproduced by permission of the Editor.

financially/economically feasible, mainly in developing countries,
but not to the extent believed in the 1970's.   The large nuclear
programmes in developed countries are likely to absorb most
investment in non-fossil fueled generation.   Fig. 1.11 shows an
estimate of the total installed and installable hydro capacity. Any
substantial development of hydro will mean a large flow of investment
from developed to developing countries, where most potential hydro
capacity is located, particularly in the period 1995 to 2015. Because
developed countries are reluctant to respond to this, overall hydro
capacity will at most quadruple between 1981 and 2020.   A good
yardstick is that 1 barrel of non-renewable oil can be substituted
by about 600 to 700 kWh of renewable hydro, if one pays an extra sum
of about US$0.10 to 0.30 per kWh for the extra investment required
for hydro plant.

1.2.9   Renewable energy production

    Before any estimates of the contribution from renewable energy can
be made some formal definition is needed on how to measure their energy

output. The exajoule is the normal unit of energy as in Table 1.1 and this can express outputs from mixed types of renewables, within which hydro, solar, geothermal, biomass, wave and wind, are likely to play the most important rôles before 2020. In Table 1.1 about 100 exajoules are shown from non-hydro renewables in 2020. This is about two-thirds of all primary energy produced by oil and gas before 1973, and about equal to the energy from oil in both 1972 and 2020; we are talking in terms of a large component of renewable energy. To accomplish this economically will be an achievement, but it is possible, because of the social, environmental and political pressures against the use of non-renewables.

1.2.10 Unconventional oil and gas

Much time, money and organisation are needed and countries must get together before the technology, economics, finance and institutional framework are available to obtain substantial outputs from sources such as tar-sands and shale-oils; but pressures on conventional oil and gas resources are likely to achieve this by 2020. Therefore, 40 EJ, almost as much as from hydro power, is shown for that year in Table 1.1 for unconventional oil and gas production.

1.2.11 Conservation

It is necessary to take a view on the likely reduction in total extrapolated energy demand achievable by conservation before drawing up any world projected supply/demand position on energy. The author believes at least a 50% reduction in presently extrapolated world primary energy demands can be realised by 2020; that 60% reduction is not untoward if full use is made of fuel substitution and of increased overall energy efficiency.

1.2.12 Energy demand

Table 1.6 Extrapolation of historical demand trends
($1EJ = 1$ exajoule $= 10^{18}$ joules)*

During the period	Annual world energy consumption increased by (%)	Which when extrapolated to 2020 corresponds to (EJ)
1860-1925	2.0	700
1860-1975	2.6	915
1925-1975	3.3	1306
1933-1975	4.1	1918
1960-1975	4.3	2111

* ibid., Table 6. Reproduced by permission of the Editor.

Table 1.6 shows the result of extrapolating past energy demands into the future. Although the resulting range is broad, it indicates that the demand for primary energy in 2020 is likely to be within the range 700 to 2100 EJ.

If we use the 50% reduction in extrapolated primary energy demands due to conservation mentioned above we arrive at about 1000 EJ for world energy demand by 2020; a 60% reduction would give 850 EJ. We can back-check to some extent because at the heart of the rate of growth of energy demand lies the coefficient of 'energy elasticity', or the 'energy coefficient'. Roughly, this coefficient is the percentage rate by which energy must grow per annum to sustain a 1% per annum increase in economic activity. This 'energy coefficient' is below unity for most developed and above unity for most developing countries; as economies mature the 'energy coeficient' decreases. The question is, what decrease will take place as countries mature in the period up to 2020? We steer a middle course in so far as improvement in the energy coefficient is concerned. The end product of our endeavours, the extrapolated figures shown for energy demand/supply in Table 1.7 and Fig. 1.12, are broadly in line with

Table 1.7   Potential world primary energy production and demand (1 exajoule $= 10^{18}$ joules)*

Resource	1972	1985	2000	2020
Coal	66	115	170	259
Oil	115	216	195	106
Gas	46	77	143	125
Unconventional oil and gas	0	0	4	40
Nuclear	2	23	88	314
Hydraulic	14	24	34	56
Renewable, solar, geothermal, biomass	26	33	56	100
Total	269	488	690	1000
Demand	269	363	570	1000

1EJ =   1 quadrillion Btu
        1 trillion cubic feet of natural gas
       34 million tonnes of coal
      159 million barrels of oil

all the above reasoning and assumptions.   Figure 1.12 shows the

KIELY, J.: 'World energy in the 21st century', Chartered Mechanical Engineer, May 1980, Table 8. Reproduced by permission of the Editor

overall demand in relation to supply and also shows the individual demand curves of groups of economies.

Table 1.7 and Fig. 1.12 both show a 'cliff-hanger' situation for our 'ultimate' year of 2020 when the potential supply of world primary energy is shown to just match the forecast demand. This result was partly inevitable because we chose 2020 as our 'ultimate' year for making projections of energy production; yet this seems a commendable 'mean' position to take here because it assumes neither a complete removal from nuclear programmes of all constraints nor that such constraints will be as great as those experienced since 1976/77. The nuclear contribution of Fig. 1.12 is the component most likely to be at risk.

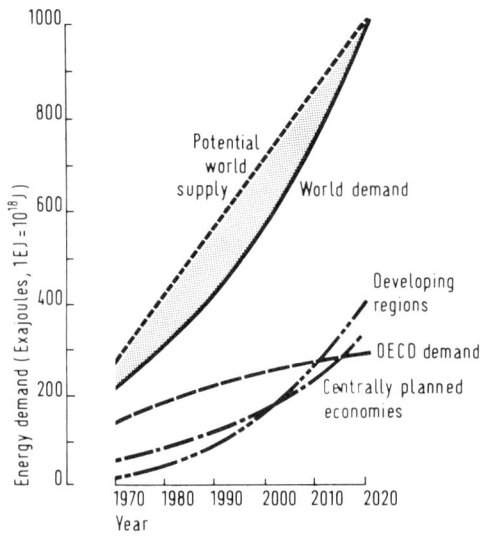

Fig. 1.12   World energy supply and demand

KIELY, J.: 'World energy in the 21st century', Chartered Mechanical Engineer, May 1980, Fig. 5. Reproduced by permission of the Editor

However, the small surplus up to 2020 shown in Table 1.7 does not mean that all is necessarily well on the energy scene at 2020. Table 1.7 is in 'single node' form; i.e. it does not allow for the physical separation of energy sources from energy uses, which must restrict the amount of energy actually available and usable at any time. However, taken by and large, Table 1.7 shows an acceptable world energy framework for use when making decisions about investment in new plant in the energy sector. Beyond 2020 things do not look too good, but by then the breeder reactor might be 'going strong', a lot of commercial energy might be coming from renewable sources, and some energy may even be coming from nuclear fusion.

REFERENCES

[1] For a fuller introduction to energy resources, sources and uses and further references, see ION, D.C.. 'Availability of world energy resources', 2nd edn., (Graham & Trotman, London, 1980); also ION, D.C.: 'World energy resources, sources and uses'. Paper given at the International Symposium on Electricity Economics and Load Management 24-28th March 1980. This chapter owes a lot to the latter paper and discussions with Mr. Ion subsequent to its presentation at the Symposium. See also TEMPEST, P. (Ed.): 'International energy options' (Graham & Trotman, London, 1981)

[2] For very good summaries of renewable and non-commercial energy see the following issues of 'UN Development Forum Business Edition', Geneva, Switzerland: No. 75, 31 March 1981; No. 76, 17 April 1981

[3] CHADENET, B., and ROVANI, Y.:'Financing of energy investments, capital and policy requirements of developing countries', Rev. d'Energie, Aug.-Sept. 1980

[4] 'Survey of energy resources 1980'. World Energy Conference, London

[5] 'Coal development potential and prospects in the developing countries'. World Bank, Oct. 1979

[6] For the general reader two good references on inexhaustible fuels are: SORENSON, B.: 'Renewable energy'. Niels Bohr Institute, Copenhagen, (Academic Press, 1979); and 'Book Review', New Scientist, 17 April 1980

[7] BERRIE, T.W.: 'Energy in developing countries'. Paper to International Symposium on Electricity Economics and Load Management, Imperial College, London, 24-28 March 1980

[8] 'A program to accelerate petroleum production in the developing countries'. World Bank, Jan. 1979; and 'Energy in the developing countries'. World Bank, 1980

[9] A concise collection of future world energy scenarios particularly suited to the objectives of this book was given in the Fourth Wilson Campbell Memorial Lecture concerning the work of the Conservation Commission of the World Energy Conference, as reported by KIELY, J.: 'World energy in the 21st century'. Chartered Mech. Eng., May 1980. The author is indebted to this article and to the papers of the 11th World Energy Conference, particularly 'Survey of Energy Resources

1980' (World Energy Conference, London), for a good deal of the reasoning in the second part of this chapter. The figures quoted are taken from the first reference because of their suitability of presentation for the objectives of this chapter; but see also TEMPEST, P.(Ed.): 'International energy options' (Graham & Trotman, 1981)

[10] 'World Development Reports'. World Bank, Aug. 1980, Aug. 1981; 'Energy planning in developing countries', Finance & Development, Dec. 1980, 17 (4). Both these references are good sources of information against which to examine the scenarios in this chapter. See also FOSTER, J.: 'Balancing energy supply and demand', Energy J., July 1981, 2 (3)

[11] BERRIE, T.W.: 'Multinational companies in the development of nuclear power', Multinational Business, (London), Apr. 1981

[12] MANNE, A.: 'Waiting for the breeder'. International Institute of Systems Analysis, Vienna, 1977

Chapter 2

The power system investment reliability and pricing rules

2.1 INTRODUCTION

2.1.1 The power sector

Electricity is the only form of energy in use today likely to be in greater demand in the next century and there may well be an 'all electric' world sometime in the future. Therefore, the power sector is one of the vital sectors to 'get right', and its importance is underlined by the urgent need to stop using exhaustible fossil fuels and start using inexhaustible fuels, many of which have electricity as an intermediate product, e.g. nuclear fuels, hydro, wave, tide, solar, wind and certain biomass.

The industrialised world is presently preoccupied excessively with the availability and price of fossil fuels and the obstacles against using the expected substitute for fossil fuels, viz. nuclear fuels.

Developing countries are concerned about the money required for investment and fuel. Investment shortage is so acute they often decide to spend less today even if it means spending more tomorrow, thus putting off the cash crisis a little longer. Developing countries have learned from bitter experience that choosing a power station for long-term reasons with lower fuel costs usually means investing much more now. Also, they need much convincing before they will commit any investment at all. This means (i) examining very carefully the trade-off between extra monies spent today and less monies spent tomorrow, bearing in mind future uncertainties; and (ii) being sure of an adequate return on investment.

Despite greater wealth, developed countries must apply the same type of tests because developed countries use the lion's share of world resources and thus bear the greatest responsibility to humanity for spending wisely.

The above tests are at the heart of power system economics.[1]

2.1.2 Power system economics

It is easiest to start by saying what is <u>not</u> included in power system economics:

(i)   System electrotechnical behaviour, e.g. voltage drops, load flows, short-circuits, stability
(ii)   Detailed utility finances, e.g. annual financial statements, financing plans
(iii) Utility management and organisation.

What remains is power-system planning, investment and pricing. To gain perspective we first look briefly at history.

2.1.3 Early history

Up to the 1920's power systems were neither interconnected nor publicly owned. What preoccupied their managers was giving satisfactory supplies to their consumers and making profits for owners; power-system economics was not required.

Amalgamations of power systems both physically and institutionally started about 1930; by about 1965 most power systems in Europe and Japan were interconnected. For the first time there could be effective planning and operation of the power sector as a whole and a need was quickly seen for power-system economics. Its use brought an immediate pay-off in the reduced size of generating-plant development programmes and the lowering of total system running costs. No one challenged spending what now seem comparatively large sums on the power sector; such challenges are comparatively recent, coinciding with people becoming increasingly vocal about, first, the needs of other public services, and secondly, the need for capital-intensive nuclear power.

In developing countries amalgamation of power systems started much later and is still continuing. In some developed countries, e.g. USA, limited (tie-line), not full, interconnection is normal. In both circumstances power-system economics has much scope.

2.1.4 The broader view

Before the amalgamations those planning and operating the systems were both the same people and exclusively engineers; economists only came in the 1960's.

Engineers in France began publishing on power-system economics in the 1930's; after World War II, French and British system planning engineers developed the subject much further. The French and Scandinavians invented methods of dealing with mixed hydro-thermal systems; by the 1960's an extensive literature existed.[2] The literature was expanded in the 1960's by developing countries, e.g. Mexico,[3] India and Jordan.

The 1970's saw a world-wide move towards tariffs based on marginal costs.[4] Today economists work alongside engineers in utilities, consultancies, government departments and in lending agencies like the World Bank. Recently project engineers and economists have joined commercial banks.

2.1.5 The subjects covered

The subjects covered in the early days of power-system economics we would now describe under 'costing'; i.e. determining the cash in current terms required to fuel and operate the system from day to day, plus the capital costs for new plant required to meet increased demand and replace old plant. Even before power-system amalgamations it was realised that a trade-off could usually be made between committing more than the minimum capital cost of any new plant in order to save in its running costs, e.g. manning, maintenance and fuel. Interconnection of power systems made possible a further trade-off by ensuring that, at any point in time, only the generating plant with the lowest running costs per kWh generated is operating.

Engineers still tend to have this 'minimum cost' type of approach to power-system economics. Economists describe it as working only on the 'supply' side; they draw proper attention to the 'demand' side, i.e. the <u>outcome</u> or benefits from investing in a project. For the economist benefits must exceed costs before any investment should be made. Attributing costs to a new project on a power system is not really difficult and can usually be done fairly accurately. Attributing benefits is neither as simple nor as accurate. It is easy to attribute revenues, which must be some minimum measure of the benefits, since consumers have shown themselves 'willing-to-pay' these amounts for the use of electricity.[5] Benefits, other than revenues, whilst often more important than revenues, are difficult to quantify, e.g. the benefit of a more reliable power system or of the convenience of electricity.

2.2 OBJECTIVES OF POWER-SYSTEM INVESTMENT

2.2.1 The basic rules

The simple objectives of the 1920's and 1930's soon became insufficient and power-system economics became complex. We will tackle this complexity by describing different groupings of objectives, remembering always that this is a simplification.

We start with the basic trade-off of spending more on capital costs today to save running costs tomorrow:

BASIC TRADE-OFFS

(i)   Should we spend more money than the minimum on installations, and thus have less to spend on other equipment and items other than equipment, in order to (a) spend less on running costs, or (b) obtain greater benefits, or (c) both?

(ii) Should we defer installations, thus having more to spend today on other equipment and other things, in order to (a) build the equipment for less investment cost, or (b) spend less on running costs, or (c) obtain greater benefits, or (d) any combination of (a), (b) and (c)?

(iii) Should we defer installations for ever, thus having more to spend on other equipment and other things, but losing any benefits directly associated with the equipment not built?

The concept of spending on one item instead of on another means giving that expenditure an 'opportunity cost' since it could have been used for 'another' purpose for which the opportunity is then lost.

At this stage we simplify the above three questions by amalgamating them into a single question which is more easily answerable, and adding two more directly related questions:

THREE BASIC RULES

(a)   How much, if anything, should we spend, and when, on a capital item? The answer is given by using The Investment Rule

(b)   How much should we charge for the output from the capital item in order to get satisfactory benefits and optimise patterns of electricity demand to suit optimum investment? The answer is given by The Pricing Rule

(c)   To what standard of performance should we build the capital item to make for optimum investment and optimum pricing? The answer is given by The Reliability Rule

---

The power-system economics in this book is mainly framed around these three rules. However, we will also deal here with other groupings of objectives, at least in so far as they concern us.

2.2.2 Different viewpoints

We can look at the power sector from many points of view, the three most important being:

(i)   the national economy, i.e. the government
(ii) the power utility, i.e. the utility directors and managers
(iii) the consumer, i.e. the customer.

Power-system economics is mostly concerned with viewpoint (i), because the sector is often owned by and always heavily regulated by government, and because electricity is sold under monopoly conditions. Also in practice it is comparatively easy to optimise with respect to (i) using (ii) and (iii) as constraints, rather than another way round.

2.2.3 National economic viewpoint

Even though we clearly define our viewpoint, it is still not meaningful to say that electricity, from that viewpoint, must be produced 'efficiently', or 'at least cost'. We must define still more closely what we mean. Though some of these adverbs are usually mentioned in electricity legislation, nowhere has the author found the adverbs defined in that legislation.
We first take our national economic viewpoint a little further and broadly relate it to two main objectives:

(i)   Increasing national aggregate economic consumption; this can be described here as achieving 'economic efficiency'; and/or
(ii) Obtaining a more equal distribution of income.

We shall look at each of these in turn, mainly as they appertain to the power sector and then concentrate throughout the book on the 'economic efficiency' objective as defined in (i) above.

2.2.4 Increasing aggregate consumption

In plain language this objective means improving the general standard of living in a country, i.e. increasing the per capita income or consumption. In order to use this objective we must first ask seemingly elementary questions on how to do any necessary arithmetic:

(i)   Can we add together directly consumption levels of different classes of consumer? If 1 kWh of electrical energy consumed by a rich man cannot be treated exactly the same as 1 kWh by a poor man, then some weighting system must be used.
(ii) Can units of consumption at different points in time be added together directly? If there is a time value of money then can a formula be devised to take it into account? This is done

in practice by a 'discounted cash flow' type of approach.[6]

(iii) Can units of consumption of different kinds of goods and services be added directly? How can we add 1 kWh of electricity to 1 kg of food? This is a problem for inter-sectoral comparison which fortunately does not concern us here.

To show how power-sector investments contribute to increasing aggregate national consumption we rank investments in terms of their total net present value to the national economy over the economic life of the project invested in, after duly allowing for the time value of money and the other weightings described above. The ranking is a measure of the excess of total discounted benefits over total discounted costs to the economy, both measured in the same unit of national aggregate consumption. All investments with a negative or zero total net present value must be rejected as they either take away or do not add to national wealth. All those investments with a positive net present value should, all other things being satisfactory, be proceeded with.

2.2.5 Net present cost

In the power sector the net present-value calculation can usually be simplified to include only the costs, i.e. to ignore benefits, because it has always been the custom to assume that meeting the electricity load forecast in full at a given standard of reliability is obligatory; another way of saying that the benefits for each alternative project will be the same. In such circumstances the projects should be ranked in order of ascending total present cost; a project with the lowest total present cost gives the greatest scope for increasing national wealth because it enables the largest amount of resources to be made available for other consumption-producing uses elsewhere in the economy.

This 'present-cost-minimisation' process is now used extensively and is often called 'finding the least-cost solution'. Care must be taken to ensure that the least-cost solution is determined with respect to (i) the total power system and not just part of it,[7] and (ii) the national economy and not just the power sector.

To get over deficiencies in the above method caused by assuming that benefits from every project invested in are the same it has become customary to check that the economic return on the investment for the least-cost solution is satisfactory. If the economic return is equal to that obtained from similar ongoing investments accepted as yielding a 'satisfactory' return, then the least-cost solution is acceptable.

Least-cost solutions and economic returns are tackled in later Chapters.

2.2.6 More equal distribution of income

We return to the second category of national economic objective. There are many ways of redistributing income, most of which do not concern us here; what we are interested in is being able to select from a number of alternative power projects that project which gives the optimal amount of income redistribution.   For example, a government can select a project according to its location, e.g. locating a project in a depressed area tends towards a more equal distribution of income.   Again, a government might choose a labour-intensive project for the same reason.

In order to use the criterion of optimal income redistribution we need to know how the benefits and cost attributable to the different projects will be distributed between differing income levels. This is not easy to find out, but when we have done so, we can attach weights[8] to all benefits and costs to allow for income redistribution effects, and proceed with the arithmetic as before to determine the least-cost solution and the economic return.

2.2.7 Advice on methods

It is appropriate to give advice here to system planners when choosing a method to use from among the many on offer.

MANUAL ON METHODS FOR POWER SYSTEM PLANNERS

When choosing a method to use, especially if you are a new-comer to power-system economics:

1   Choose as simple a method as possible consistent with what you want to do and the accuracy of your data.[9]

2   Check that the particular method chosen is recognised as 'orthodox' by the best practising power-system planners of today. Ask one or two; study their works; be as catholic as possible; examine official booklets.[10]

3   Once you are sure you have chosen an 'orthodox' method, then make sure you fully understand:
    (a)   how to use it
    (b)   the detailed logic upon which it is based
    (c)   the situations for which it has been designed
    (d)   the degree of accuracy needed in input data
    (e)   the degree of accuracy expected in output data.

4   When you fully understand the method use it properly, consistently, and as far as possible without bias, not looking too much at the results whilst in the middle. Most abuses of the methodologies of power-system economics come from such things as:
    (a)   gross misuse of an orthodox method
    (b)   inadequate understanding of any method
    (c)   using any method beyond its capacity.

2.3   PRICING OBJECTIVES

2.3.1   Introduction

A cursory glance at the types and ranges of power-sector pricing presently in use shows that these must be the end-product of many different thought processes and data bases. For example, pricing rules can be set in relation to:

> Historic costs, in accordance with the financial accounts of the utility; or
> Some estimate of future financial costs, e.g. replacement costs; or
> Costs not necessarily related directly to the financial accounts of the utility, e.g. marginal costs (see later).

Again, tariffs can be set artificially uniform for:

> All consumers of one class irrespective of geography, e.g. domestic consumers;
> All consumers supplied at the same voltage, e.g. at low voltage;
> Certain time periods, e.g. stipulated off-peak periods.

Alternatively, tariffs can be set to differentiate by:

> Geographical region; or
> Time-of-day, week, month or year; or
> Level of maximum demand.

2.3.2   The pricing rule

Although not fully accepted, most power-system planners agree that the demand for electricity can be sensitive to price.[11] A low electricity tariff encourages the use of electricity, drawing investment and fuel into the power sector, possibly to an unwarranted extent. Low or high electricity prices will affect the financial performance of the utility and the redistribution of income; for example, tariffs could be set so that large consumers, assumed to be rich, will subsidise small consumers assumed to be poor. Again, a tariff level set only for optimising national objectives may prevent the power utility from achieving its financial target.

We normally end up with a compromise between the objectives of:

(i)   Economic efficiency; i.e. optimally allocating national economic resources to the power sector within the national plan for achieving optimal growth of wealth.

(ii)   Income redistribution, i.e. optimally redistributing national income by means of the power sector.

(iii)   Achieving the financial obligations of the power utility.

We have already seen that items (i) and (ii) form a major part of the optimum investment rule, especially item (1). We now see that these items also form a major part of the pricing rule. As was the case with the investment rule, in working out the pricing rule it is normal to start with objective (i) above, treating any other objectives as constraints. In the past many tariffs were based only on objective (iii), using as their data the financial projections from the utility's historical accounts.

2.3.3 Marginal cost pricing

Stated another way, objective (i) above, the 'efficiency' objective as it is normally known, means that the price paid by a consumer for an extra kW and an extra kWh of electricity must be related directly to the additional cost to the national economy of meeting the extra kW or kWh, as worked out by the investment rule. This cost is usually termed the 'marginal cost' because it refers to a demand for electricity 'at the margin', i.e. beyond what is presently being planned to be supplied at the given level of reliability.

The divergence of any tariff from marginal costs depends not only on the distortions introduced by objectives (ii) and (iii) acting as constraints, but also on other balances within the energy sector as a whole, e.g. the relationship between the commercial price and the marginal cost of fuels which can substitute for electricity and vice versa. However, marginal-cost pricing is always the basic means of giving the 'correct' signal to consumers because it makes the optimum use of resources under the investment rule. In this respect, the pricing and the investment rules are but different aspects of the same thing. We should also take note that marginal-cost tariffs are forward looking, the 'old-fashioned' tariffs based upon historic financial accounts of the utility being backward looking.

2.3.4 Time-related tariffs

Before the introduction of marginal-cost tariffs it had long been customary to grant favours in the form of lower tariffs to consumers who avoided taking electrical energy at peak demand time. Times of peak demand are times when there is a large price to be paid both with respect to extra maximum demand (kW) and extra electrical energy (kWh). At peak demand times the most expensive generating plant to operate will be running and any extra demand at that time will require investment in new plant. On the other hand, off-peak demands require less expensive plant to operate and no investment costs for extra demands at that time. (The above is but a restatement of marginal-cost pricing.) When dealing with time-related tariffs, whether marginal-cost derived or not, it is always necessary to do some averaging.[12]

2.3.5 Practical constraints on tariffs

To be practical all electricity tariffs must pay attention to factors other than marginal costs under the pricing rule:

CONSTRAINTS ON TARIFFS

(i)   Understanding by the consumer; complex tariffs can only be used for large industrial/commercial consumers

(ii)   Metering costs; the same as (i)

(iii) Price distortion; because the national economic prices differ from the money prices due to taxes, subsidies, economic and social considerations. If the extent of any distortion is known, theoretically one can take it into account when setting tariffs. A common distortion is that between economic and money costs for fuels

(iv)   Environmental factors[13]

(v)   Income distribution, see above

(vi)   Financial objectives of the utility; we also discussed this above. A further problem may arise with prices set according to the pricing rule because a utility's financial objectives must be met annually, whereas the pricing rule follows the investment rule in looking at the longer term.

We return to pricing in Chapter 9.

REFERENCES

[1]   BERRIE, T.W.: 'It's always jam tomorrow', Elect. Rev., 202, (5),
       3 Feb. 1978

[2]   One of the best bibliographies on this subject is given after
       Chapter 13 in: TURVEY, R., and ANDERSON, D.: 'Electricity
       economics'. A World Bank Research Publication (Johns Hopkins
       University Press, Baltimore, USA, 1977)

[3]   FERNANDEZ, G.: 'Electric power planning' in: GOREUX, L., and
       MANNE, A.(Eds.): 'Multi-level planning case studies in Mexico'
       (North Holland, 1973)

[4]   'Symposium on peak load pricing', Bell J. Economics, 1976, 7,
       pp. 197-250.   TURVEY, R., and ANDERSON, D.: 'Electricity
       economics', ibid.   CREW, M.A., and KLEINDORFER, P.R.:
       'Reliability and public utility pricing', American Economic
       Rev., Mar. 1978, 68.   'Second best pricing with stochastic
       demand', ibid.   'Marginal costing and pricing of electrical
       energy'.   Proceedings of the State of the Art Conference,
       Canadian Electrical Association, Montreal, May 1978.
       MUNASINGHE, M.: 'Economics of power system reliability and
       planning' (Johns Hopkins Press, Baltimore, USA, 1979). ROZALI,
       B.M.A., BERRIE, T.W., and MURGATROYD, W.: 'Planning the develop-
       ment of a power system'. Elect. Rev., 2 Nov. 1979, 205, (17)

[5]   TURVEY, R., and ANDERSON, D.: 'Electricity economics', loc.
       cit., Chap 11

[6]   GITTINGER, J.P.: 'Compounding and discounting tables'. Pub-
       lished for the World Bank (Johns Hopkins University Press,
       Baltimore, USA, 1973)

[7]   BERRIE, T.W.: 'The economics of system planning in bulk
       electricity supply', Elect. Rev., 29 Sept. 1967

[8]   SQUIRE, L., and VAN DER TAK, H.G.: 'Economic analyses of
       projects'. Published for the World Bank (Johns Hopkins
       University Press, Baltimore and London, 1975)

[9]   BERRIE, T.W.: 'Economics is beautiful'. Commentary, Elect.
       Rev., 1978, 202, (9), p. 21

[10]   Nowadays there are some very good booklets on project economics
       and appraisal.   They are published by many international
       organisations such as the World Bank, Washington DC, USA; UNIDO,
       Vienna, Austria; and OECD, Paris, France. The reader who wishes
       to obtain a copy of any of these is advised to write to these

organisations with the simple addresses given above; but see especially LITTLE, I.M.D., and MIRRLEES, J.A.: 'Project appraisal and planning in developing countries' (Basic Books, New York, USA)

[11] 'Report on the Working Group on energy elasticities'. Energy Paper 17 (HM Stationery Office, London, 1977)

[12] TURVEY, R., and ANDERSON, D.:'Electricity economics', loc. cit., pp. 338-347; and 'Power sector planning manual'. Overseas Development Administration, UK, June 1979, p. 80

[13] 'The polluter-pays principle'. OECD Publication, 1975. HIGHTON, N., and WEBB, M.:'Sulphur dioxide from electricity generation: Policy options for pollution control', Energy Policy, Mar. 1980

Chapter 3

Reliability standards

3.1 THE RELIABILITY RULE

3.1.1 Introduction

Chapter 2 identified three basic rules of power system economics; i.e. the Investment Rule, the Pricing Rule and the Reliability Rule.[1] This chapter deals with the Reliability Rule. To get the perspective right we again note that investment in the power sector in the decade 1981 to 1990 will be about US$400 billion in the USA and about US$450 billion in the total developing world.[2] This massive capital investment in electricity will get larger as rising fossil fuel prices force more investment in expensive-to-build nuclear and hydro stations, together with novel renewable forms of energy.

Most systems are designed to have a margin of plant above that required to meet the expected maximum demand in order to cater for planning and operating contingencies; if this margin is even 1% higher than needed then much capital expenditure is wasted; if this margin is set too low the economy and consumers are at risk. Because the sector's capital is so large we must know what is the optimum plant margin.

3.1.2 Optimum reliability standard

Planners are used to working out the total (capital plus running) system costs for different levels of reliability,[3] but most now realise that this alone is not enough.[4] Their difficulty is to decide on a reliability level, having found the system cost for different levels. To find the optimum level we must know what the consumers/economy lose by 'brown-outs' and 'black-outs'.

Most presently planned reliability levels are based on past experience.[5] To take one further step is theoretically easy but practically difficult[6] because we are concerned with supply quality and not quantity, and effects of changes in quality are more difficult to measure than changes in quantity. Also, these changes can be gradual, e.g. changes in frequency and voltage in 'brown-outs', or sudden, e.g. total system failures in 'black-outs'. Two types of cost are involved:

(i)   Costs incurred (saved) by installing more (less) plant capacity to improve (worsen) the supply quality. These are the total system costs mentioned earlier (SC).

(ii) Costs saved (incurred) by the consumer/economy because of a better (worse) supply quality. These are outage costs (OC).

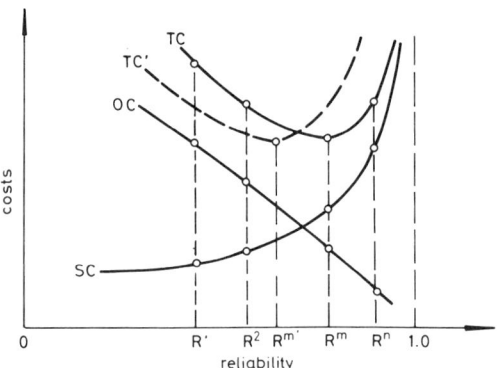

Fig. 3.1   Outage costs, supply costs and total costs as functions of the reliability level

> MUNASINGHE, M. 'Power system economics, pricing and reliability standards: Part II: Optimal reliability and system planning'. Paper presented to the International Symposium on Electricity Economics and Load Management, Imperial College, London, 24-28 March 1980. Reproduced by permission of Imperial College, London, Power Systems Group.

The supply quality, i.e. the reliability level (R), should be increased until the incremental outage costs (OC) are just equal to the incremental system costs incurred (SC) in Fig. 3.1. We are then at the optimum reliability level.

3.1.3   Costs incurred

The rising curve (SC) in Fig. 3.1 is easily and often drawn. On most power systems, increasing the amount of plant to increase the reliability will mean either:

Bringing forward planned plant; and/or
Installing additional, often base-load plant; and/or
Installing specialised, often peak-load plant; and/or
Keeping longer in service plant otherwise to be scrapped.

The total (capital plus running) cost changes to the whole system of doing any combination of these alternatives can be worked out on a system optimisation model (see Chapter 7). The least-cost solution is thus found for each point on curve (SC) from:

Total discounted costs to the system of meeting a given demand forecast at least-cost using improved reliability of supply;

less

Total discounted costs to the system of meeting the same demand forecast at least-cost using the initial reliability standard.

3.1.4 Outage costs

Drawing the falling curve (OC) in Fig. 3.1 is much more difficult. Most figures will be empirical, measured for each consumer class and totalled in some logical way. For industrial and commercial consumers outage costs saved by an increased reliability of supply are lost output to the economy plus any inconvenience to the consumer's well-being, the economist's 'consumer surplus'. For residential consumers one way of determining lost consumer well-being is by valuing lost leisure activities, e.g. listening to the radio, watching TV or reading; plus inconveniences caused by lost heating/air-conditioning or cooking facilities.[7]

3.1.5 A pragmatic approach

Because measuring outage costs is difficult it is useful to turn this problem round.[8] We illustrate this by a simple example. Assume a system has an optimal plant-mix by fuel type and generator size and that gas turbines cost £200/kW with an economic life of 30 years. If the discount rate is 10% the annual capital charges on the investment in a gas turbine for running at system peak are £21.2 per kW.[9] If one 'black-out' affects 5% of consumers and the number of hours of lost supply per consumer per year because of generation shortage is 1.0, then the expected number of hours lost per consumer per year for a given number of hours of 'black-out' is

$$\sum_{i=1}^{n} L_i \, p(L_i)$$

Where

L_i = total number of hours of black-out in any year for consumer i

$p(L_i)$ = probability of consumer i being affected.

n = total number of consumers.

We build on this formula; taking the above figure of 5% of consumers affected by a 'black-out', then $p(L_i) = 1/20$. If L_i is 20 hours, then: Number of hours lost per consumer per year is $(20 \times 1/20) + (0 \times 19/20) = 1$, as previously assumed. Since the marginal cost of supply is £21.2/kW p.a. and the duration of 'black-outs' per year is 20 hours, then:

Implicit marginal value of electricity at existing security standard is £21.2 divided by 20, i.e. £1.06/kWh.

This method can be taken one step further before losing its simplicity; i.e. to calculate the implicit marginal value of electricity measured in terms of outage costs averted. Suppose we plan to improve our reliability standard by adding plant at a cost of £21.2/kW p.a. as above. We want to reduce the expected duration of black-out per consumer from 1.0 hour, as above, to (say) 0.5 hour, leaving the proportion of consumers affected unchanged. Using the figure of 0.5 hour, the expected duration of supply interruptions would be 10 hours per year, and then:

Implicit marginal value of electricity is now £21.2 divided by 10, which is £2.12/kWh.

Repeating the calculation for a range of security standards gives Fig. 3.2.

Using Fig. 3.2, we compare the present implicit marginal value of electricity given by the curve DC with the value-added per kWh for industrial and commmercial consumers to arrive at some idea as to whether the existing quality of supply is about 'right'. If we can add the leisure value of electricity to residential consumers the method's value increases (see later).

There is an urgent need for using this type of method on existing systems. In the past, standards of reliability have tended to be set high rather than low; planners have a fear of plant shortage. However, too much money is at stake to allow anyone to indulge in such fears if they are unwarranted.

The framework against which to evaluate both outage costs and the more familiar system costs is economic. Optimisation in the power sector should be basically from the point of view of the economy rather than from the points of view of either the utility or the consumer (see Chapter 2).

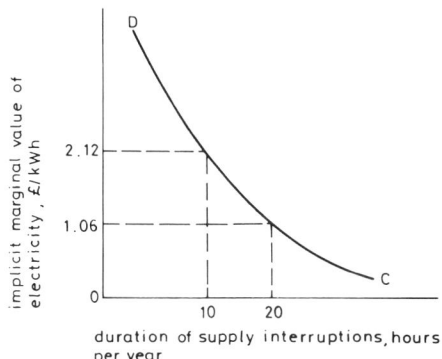

Fig. 3.2   Implicit marginal value of electricity

WEBB, M.: 'Power system economics, reliability standards
and pricing'. Paper given at the International Symposium
on Electricity Economics and Load Management, Imperial
College, London, 24-28 March 1980.   Reproduced by permission of Imperial College, Power Systems Group.

3.2   PRACTICAL APPROACHES

3.2.1   A practical methodology and model

Fig. 3.3 shows the flow chart of one practical methodology for
optimising the reliability standard.   It should be used something
like this:

First, develop a set of models to analyse the outage costs incurred
by different consumer classes due to electric power shortages of
different intensities and durations (box 3 in Fig. 3.3).

Concurrently, make a load forecast for, say, 25 or 30 years ahead
disaggregated by consumer classes, based on present or a predetermined evolution of electricity prices (box 2 in Fig. 3.3).
Then, prepare several alternative, least-cost system development
programmes to meet the forecast at several levels of reliability (box
1 in Fig. 3.3).

Finally, estimate for the entire forecast period the expected
reliability for each development programme in terms of frequency of
brown-outs and black-outs and the duration of same, times of
occurrence, and the number of consumers affected (boxes 4 and 5 in
Fig. 3.3 giving SC and OC, respectively).

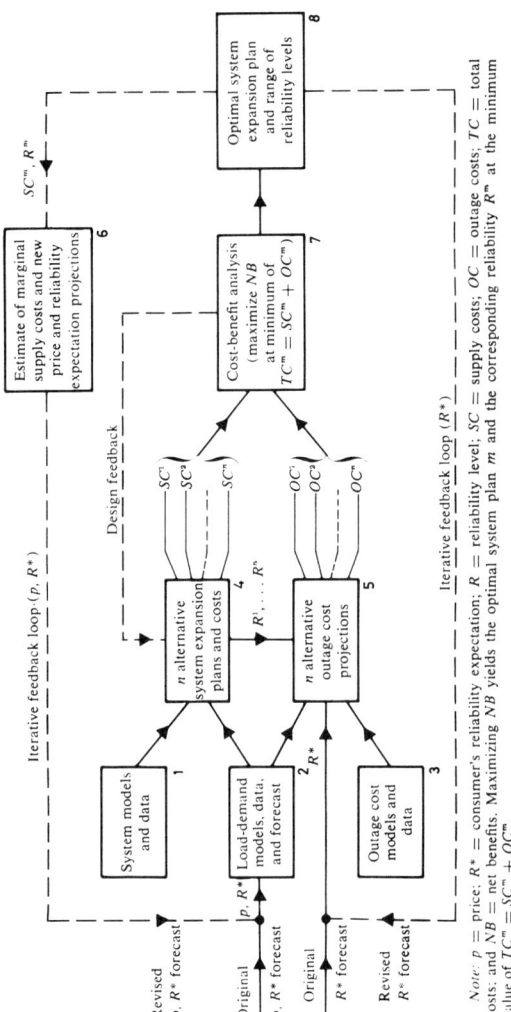

Fig. 3.3 Flowchart for the implementation of the methodology

Note: p=price; R*=consumer's reliability expectation; R=reliability level; SC=supply costs; OC=outage costs; TC=total costs; and NB=net benefits. Maximising NB yields the optimal system plan m and the corresponding reliability R^m at the minimum value of $TC^m = SC^m + OC^m$.

MUNASINGHE, M.: 'Power system economics, pricing and reliability standards; Part II: Optimal reliability and system planning', presented at the International Symposium on Electricity Economics and Load Management, Imperial College, London, 24–28 March 1980. Reproduced by permission of Imperial College and Johns Hopkins University Press; it appears also in Reference 23.

We can now return to the outage cost averted for consumers if
reliability is improved. By substituting the difference in expected
brown-out and black-out frequency and duration between least-cost
development programmes in the outage cost models described above we
can trade-off extra system costs (SC) spent in increasing capacity
in a least-cost way against outage costs (OC) averted by installing
the increased capacity. Our optimum least-cost development programme
is that for which $\Delta OC = \Delta SC$ (see Fig. 3.1). This means we choose the
optimum long-run least-cost development programme which minimises
total costs (TC) in Fig. 3.1, where

$$TC = SC + OC$$

provided we can make evaluations sufficiently accurately (box 7 of
Fig. 3.3).

3.2.2   Feedback in the model

One strong feedback in the model is via the impact of prices on
the load forecast. This may require further iterations through the
model if the newly optimised system programme requires changes in
the prices originally assumed for the load forecast. We can also
take into account the influence of changes in reliability
expectations of consumers on both the load forecast and the outage
cost saved. We can iterate through the model enough times to arrive
at a mutually self-consistent set of price, demand and optimum
reliability levels. Provided prices are basically set within the
framework of long-run marginal costs of supply, and because each
development programme considered is always the least-cost solution,
this self-consistency will guarantee that the Investment, Pricing
and Reliability Rules are all satisfied.

Fig. 3.4 illustrates that the above is true. Suppose the optimal
price P_0 has in the starting year been set by the Pricing Rule, in
accordance with the long-run marginal costs (LRMC) of least-cost
optimum development, i.e. the cost $LRMC(R^0)$ at the point A where
supply can be taken up by the electricity market. We derive the curve
$LRMC(R^0)$ from the system development programme with the optimal
reliability level R^0, i.e. keeping R^0 fixed. Thus initially both
price P and reliability R are jointly optimised.

Now if, after some period of time, the demand against price curve
D in Fig. 3.4 shifts from D_0 to D_1, the optimal price is now not
necessarily P_1 on the same $LRMC(R^0)$ curve. In fact, using a static
LRMC curve corresponds to the planner's 'traditional' method of
working with a fixed target reliability level R^0. The optimal
reliability level may have changed to R^1 and the appropriate curve
to use is now $LRMC(R^1)$ with the optimal price P_1. As demand increases,
the 'dynamic' optimal long-run marginal cost curve $LRMC_{DYN}$ lies along
AB. B is unlikely to be precisely definable because, although the
LRMC curves are usually well known from 'supply' side least-cost

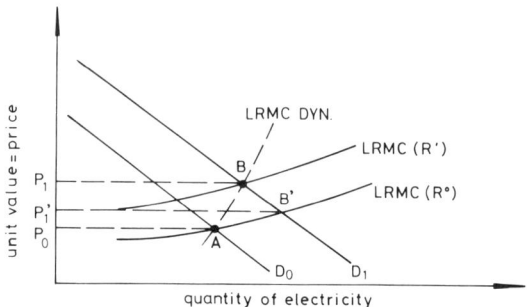

Fig. 3.4   Evolution of demand and reliability

> MUNASINGHE, M.: 'Power system economics, pricing and
> reliability standards; Part II: Optimal reliability and
> system planning'. Presented at the International Sym-
> posium on Electricity Economics and Load Management,
> Imperial College, London, 24-28 March 1980. Reproduced
> by permission of Imperial College, London, Power Systems
> Group

solution considerations, the shape and position of demand curves D_0
and D_1 will not be known very precisely unless a lot of data is
available on the 'demand' side in the form of price versus demand
for electricity.

Because in Fig. 3.4 the point B is badly defined and a shifting
target, a trial-and-error approach is necessary and often quite
practical. We examine how, as demand grows, plant is added to expand
the output potential of the system.   Price and reliability are
optimised iteratively, as described above.   The best procedure is:

• When reliability is being optimised, assume price and demand to
  be unaffected;

and

• When price is being optimised, assume reliability and demand to
  be unaffected.

Through successive iterations we then reach a mutually self-
consistent set of optimal price, reliability and demand levels at
point B in Fig. 3.4, or as near to B as the available data permit.

3.2.3 Generating-plant reliability

There is a direct relationship between total power-system reliability and component reliability. To prevent poor supply quality we provide sufficient spare generating plant capacity to meet a credible range of contingencies.[10] In operation there must be spare plant in both (i) 'spinning' reserve and (ii) 'standby' categories. Category (i) reserve is for immediate pick-up and category (ii) is for short and medium-term pick up. Such spare plant is needed mainly to cater for a generation failure. When planning ahead a plant margin must be built into the least-cost development programme to cover this and other uncertainties. It is then called the 'planning plant margin'. Because the number of possible contingencies in the future is larger than the number in the event, operating margins are less than planning margins, sometimes by quite a large amount. Although most brown-outs and black-outs per consumer are due to unreliability of the network, generation outages have a much more profound and widespread effect on the system. Most reliability analysis is thus done with respect to generating plant. Network reliability cannot be easily tackled on a 'margin' basis; it can only be tackled properly on a probability basis (see below).

3.2.4 Measures of reliability

In the light of the basically probabilistic character of system reliability coupled with component reliability, it seems advisable to use indices of reliability for the system as a whole which can be coupled with component outages by combinational probabilistic methods. It is possible, using these methods, to compute the long-run probability that a large complex system will be in a given reliability state; for example, able to meet the demand for electricity in full, or failing to meet the demand by any given margin. This is done by using probabilities on power-system loading in conjunction with information regarding the probable availability of individual system components, the information being combined according to the normal axioms of probability.[11] The most familiar indices of reliability as described below are calculated in this way.

A sound approach to reliability is based on overlaying probabilistic techniques with stochastic processes evolving over time. The system is modelled as a discrete-state, continuous-transition Markov process.[12] At any moment in time the system may change from one state to another due to events such as component outages or actions designed to restore normal operation. Once all possible states of the system have been identified, together with the rates of transition between these states as determined by the corresponding contingencies or restorative control actions, then the system can be fully analysed. The transition rate matrix, which specifies the transition rates between all possible pairs of states, provides a useful mathematical summary of the Markov process.

We now discuss the main probabilistic indices of system reliability.

3.2.5 Reliability indices

The 'loss-of-load-probability' (LOLP) index is probably best known. It is sometimes called the 'loss-of-load-expectations' index. LOLP interprets system failure in terms of inability to meet the daily peak demand. In generation studies LOLP is broadly the average number of days over some fairly long period during which peak demand is expected to exceed available generating capacity. The target LOLP mainly used in the United States is 1 day in 10 years, in European countries it varies from 1 day in 15 years to 1 day in 2.5 years.

In network planning, time is measured continuously rather than in terms of the daily load cycle. The LOLP is then the long-run average fraction of total time that the system is expected to fail, failure being defined as an interruption of supply, i.e. a 'black-out', although it can be defined as a 'brown-out', i.e. percentage frequency/voltage reduction, or even a component overload. The LOLP index is often supplemented by the expected-loss-of-load (XLOL) index, i.e. the expected magnitude of unsupplied load, once a failure has occurred.

Another index is the 'loss-of-energy-expectation' (LOEE), or the 'expected-unserviced-energy' index. LOEE is the expected amount of energy not supplied due to outages in the long-term expressed as a fraction of the total energy demanded.

Two more recently used indices are the expected frequency or mean recurrence time between outages, and the expected duration of such events when they do occur. These are used as a pair of reliability indices (FAD), mainly in network planning.

There are two main methods for calculating reliability indices. Analytical methods rely on the direct or analytical manipulation of basic probability axioms;[13] Monte Carlo methods use the computer for throwing imaginary dice first to simulate the various random events, e.g. mean time to next failure, repair time of components etc., and then to estimate the overall development programme reliability levels.[14]

3.2.6 Comparison of different reliability indices

Each index has strengths and weakness; none give a complete description of the system state. Although a measure of probability of system failure, LOLP does not give the expected magnitude of load lost during a failure; also, because it is a long-run concept, the impact of operating procedures and criteria designed to modify the risk of failure is ignored.

Although XLOL does provide the expected magnitude of loss, it does not provide either the probability or average duration of an outage; neither does LOEE. Like LOLP, FAD does not give the mean size of failure. However, as FAD is commonly specified at the load point, the number of affected customers and magnitude of the interrupted load can be deduced from it.

Unfortunately, the indices differ with respect to their underlying models, data interpretations, computation methods, reliability interpretation; and thus are neither comparable nor consistent. They can give different rankings of projects, and can be calculated[15] and interpreted[16] in different ways.

Nevertheless, such indices can be helpful in the formal, structural process of system planning.

3.3 HOW TO CALCULATE OUTAGE COSTS

3.3.1 Introduction

We define outage costs as 'direct' or 'indirect' according to whether they are incurred:

(i)   because an outage <u>actually</u> takes place; or
(ii) because an outage <u>might be expected</u> to take place.

During an <u>actual</u> outage direct outage costs may be incurred since normal productivity slows down or stops altogether. Indirect outage costs can be incurred without an actual outage because consumers: (a) use less efficient or more costly ways in order to be less susceptible to outage disruptions; or (b) purchase standby energy sources. Indirect outage costs cannot be attributed to any particular outage; they depend on the general perceived reliability level of supply, but they represent real resource costs to the economy, to be taken into account when comparing reliability levels of alternative development programmes, including the effects of 'load management' (see Chapter 9). One difficulty in evaluating outage costs associated with particular reliability levels is that such costs normally consist of a mix of direct and indirect components particular to a given situation at a given time.

3.3.2 Direct and indirect outage costs

Direct and indirect outage costs associated with a particular level of reliability depend in their turn on the 'expected' and 'actual' levels of reliability. We illustrate this point by considering two situations.

Situation I

In situation I consumers <u>expect</u> a low reliability level R^{L*}; they adjust their household and productive activities, or purchase alternative energy sources to reduce direct outage costs. This results in indirect outage costs as described above. However, the sum of direct and indirect outage costs also depends on the <u>actual</u> level of reliability. If this turns out to be low, direct outage costs are large despite the indirect preventative outage costs incurred by consumers. Total outage costs will thus be high. If actual reliability turns out to be high, direct outage costs will be small; total outage costs will be less than in the case of low actual reliability.

Situation II

In situation II consumers expect a high reliability level R^{H*}; they do little to adjust their behaviour. Few indirect outage costs are incurred. Total outage costs are high if actual reliability turns out to be low. Such costs will be smaller if actual reliability turns out to be high.

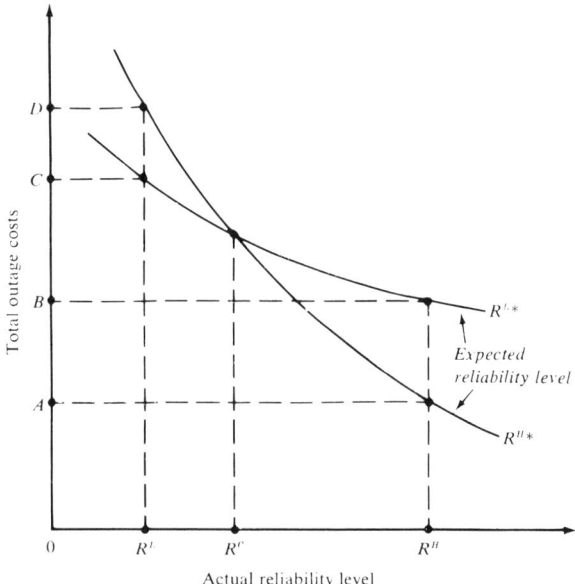

Note: R^{H} = actual high reliability level; R^{L} = actual low reliability level; and R^{c} = critical reliability level. R^{H*} = expected high reliability level, and R^{L*} = expected low reliability level.

Fig. 3.5   Relationship between total outage costs and the levels of actual and expected reliability*

The relationships between expected and actual reliability and

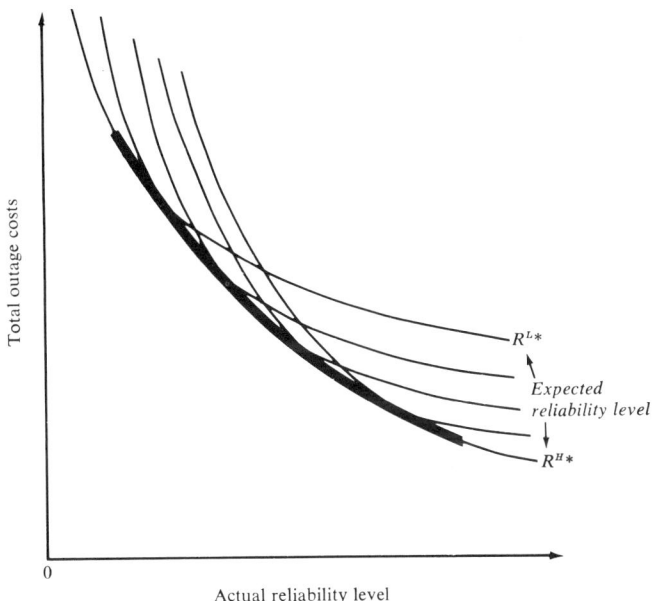

Fig. 3.6   Envelope curve of total outage costs (heavy line)*

outage costs are demonstrated graphically in Figs. 3.5 and 3.6. In these Figures each curve is a graph of total, direct plus indirect, outage costs against actual level of reliability, but shown against a given level of expected reliability. At a given level of expected reliability, outage costs are high when the difference between actual and expected reliability is large.   Outage costs decrease with increasing actual reliability for any given level of expected reliability (see Fig. 3.5). Total outage costs are highest with high expected reliability and low actual reliability R^{H*}, R^L. Lower total outage costs are with low expected and low actual reliability R^{L*}, R^L. The third lowest total outage costs are with low expected and high actual reliability R^{L*}, R^H.   Finally, total outage costs are lowest with both high expected and high actual reliability R^{H*}, R^H.

MUNASINGHE, M.: 'Power system economics, pricing and reliability standards; Part II: Optimal reliability and system planning'. Paper presented at the International Symposium on Electricity Economics and Load Management, Imperial College, London, 24-28 March 1980. Reproduced by permission of Imperial College, and Johns Hopkins University Press; it appears also in Reference 23

Reliability expectations lie between R^{L*} and R^{H*}. There will be many curves relating reliability to outage costs, one curve for each level of reliability expectation. At each level of reliability the curve for the lowest possible outage costs is that for which actual and expected reliability are equal. We thus trace an 'envelope' curve (Fig. 3.6) indicating outage costs resulting if planners and consumers could correctly both anticipate the actual reliability and then take steps to minimise the resulting outage costs. Drawing the envelope from time to time is useful for load management (see Chapter 9).

It is difficult to determine how changes in actual reliability affect consumers' expectations of reliability; we estimate outage costs assuming that consumers' expectations remain unchanged over some range of actual reliability. With this simplification, adaptations for averting outage cost such as buying standby generators are assumed to occur only after the actual reliability falls below some critical level, which will vary for different consumers. In Fig. 3.5, if the actual reliability level is reduced from R^H to R^L, outage costs increase by AD, assuming that reliability expectations remain unchanged at R^{H*}. This will happen once the critical reliability level R^C has been passed through. In practice consumers take time to adjust and the increase in actual outage costs will lie somewhere between AC and AD. Often the easier determined outage cost AD is a good enough estimation.

3.3.3 Estimating outage costs

There are two main schools of thought on outage-costs estimation. The first estimates these costs from observed willingness-to-pay for planned electricity consumption;[17] the second estimates them in terms of the effect of outages on production of goods and services.[18]

Estimating outage costs from observed willingness-to-pay is part of a general approach on optimal pricing for utilities under conditions of uncertainty, i.e. of stochastic demand and supply. The variable to be maximised is net economic welfare, i.e. expected area under the price against demand curve (D_0 in Fig. 3.4); this area has the units of price times quantity, corresponding to the value of planned electricity consumption minus costs saved by not supplying the consumption. The willingness-to-pay approach assumes that electricity provides a direct satisfaction to consumers including the economist's 'consumer surplus'. In this approach outage costs are estimated in terms of lost consumer satisfaction with lost consumer surplus.

The second approach is in terms of the effect of outages on various types of output or production. This approach does not suffer the disadvantage of having to make many assumptions when calculating lost consumer surplus. It treats electricity as an intermediate input used to produce various goods and services. It is these goods and services which provide satisfaction, not electricity. Outage costs are then measured by the costs to the economy of outputs not produced produced because of brown-outs/black-outs. This second approach is especially to be recommended when there are serious difficulties in using willingness-to-pay.

3.3.4 Willingness-to-pay

The difficulties of using willingness-to-pay as a measure of the outage costs are:

(i) Observed willingness-to-pay for <u>planned</u> electricity consumption is not necessarily an accurate indicator of willingness-to-pay to avoid an <u>unplanned</u> outage.[19]

(ii) An unplanned outage will disrupt activities complementary with electricity consumption. Therefore actual outage costs may be greatly in excess of observed willingness-to-pay.

(iii) We measure losses of consumer surplus assuming that load shedding takes place according to a scale of willingness-to-pay. This may be neither reasonable nor practical. If it is an outage due to a distribution failure the ability to carry out load shedding in any predetermined order is severely limited. Experience has shown that actual outage costs exceed those estimated on the basis of lost consumer surplus.

(iv) The method of lost consumer surplus requires a fairly exact knowledge of the price against demand curve of various types of consumer.[20] Empirically estimating such demand curves is a problem and subject to considerable error.

3.3.5 Effect on productive activity

Given the above difficulties, outage costs are more conveniently measured in terms of the outage effects on various kinds of productive activity. In production, capital and labour combine with inputs such as raw materials and intermediate products to produce a time stream of outputs. Without market distortions, i.e. under perfect competition, the net economic benefit of a marginal unit of output in a given time period equals the economic value of that output minus the economic value of the inputs. (When market distortions are present, 'shadow pricing' of inputs and outputs is necessary to allow for differences in perception, e.g. between the viewpoints of the economy and of the consumer concerning what something is worth).

The total benefit to the economy from a stream of marginal units of outputs over time equals the total present value of the resulting stream of these discounted net economic benefits. The economic cost of having a power sector with a reliability of supply less than perfect is measured by the resulting reduction in this total present value. An outage disrupting production or commercial service reduces the net benefits from these activities. Direct outage costs are incurred because either the costs of inputs are increased or the value of outputs is reduced, or both. Some outages cause spoilage; means of production or of providing services may become idle. Spoilage leads to an economic cost equal to the value of the final product not now being made available, minus the value of inputs not used because the final product is not produced. When the output value cannot be determined, e.g. for household and public sector outputs not directly sold on the market, we must define the cost of producing the spoiled output another way. We illustrate this by an example.

3.3.6 Effect of electricity outages on residential consumers

A case study has been carried out by the World Bank to test empirically the second method of calculating outage costs.[21]   The long-range distribution plan of a city was optimised according to the procedure summarised in Fig. 3.3. Only the distribution system reliability was varied, the reliability of generation and transmission being held constant. All figures are in 1977 price levels.

During a 'black-out' residential consumers rescheduled housekeeping chores dependent on electricity without much inconvenience. Cooking was not disrupted since in that city it is done with natural gas. However, leisure activities were significantly affected by outages since leisure time in most households was restricted to a relatively fixed evening period. Electricity could be considered essential during certain evening hours for watching TV, reading, dining etc.

The results of a survey and analysis of outage costs of residential consumers confirmed what has been described more fully in other published work.[22]

(i)   The chief impact of unexpected power outages on households was the loss of a critical 90 minutes leisure period occurring during the evening when electricity was considered essential. Domestic activities interrupted during the day could be rescheduled relatively easily.

(ii)   Over this 1.5 hour critical evening period, the monetary value of lost leisure could be measured approximately in terms of the net wages or income-earning rate of affected households, as confirmed by their short-term willingness-to-pay to avoid power outages. Estimated residential outage costs were in the range of US$ .30 to US$ 2 for each kWh lost or not consumed due to outages.

(iii) The principal advantage of this method is its reliance on income data which is relatively easy to obtain.[23]

Industrial consumers in the city suffered outage costs because materials and products were spoiled, and normal production could not take place. Disrupted production has an economic cost of idle capital and labour during the outage and the restart period. With slack productive capacity some lost value-added may be recovered by using this capacity more intensively during normal working hours. Overtime schedules may make up lost production. A survey of the 20 principal industrial electricity users in the city was made to determine outage costs for power failures of various durations between 1 minute and 5 hours. The results show a wide variation: US$ 1 to US$ 7 per kWh lost, depending on type of industry, outage duration and time of day of the outage. This approach also ranked industries in terms of sensitivity to outages, useful information to the utility for emergency load shedding and load management (see Chapter 9).

Public illumination outages imposed a cost of 'foregone' community benefits, e.g. security, improved motoring safety etc. These benefits foregone are worth at least as much as the net supply cost the community would have incurred for public illumination during the outage periods, e.g. the annuitised value of capital equipment and routine maintenance expenditures. Electricity costs are not included since they are not incurred during outages.

Network expansion plans were designed, based on the principle of providing the highest reliability of supply to areas with the highest outage costs. The whole outage cost exercise described above was then repeated several times using these plans. This procedure of continuous feedback and iteration yielded an optimal power-system expansion plan. This aimed to provide a high reliability of supply in both the city centre areas which had a high population density, and in the main industrial zone. Other areas were served at a lower reliability level; but the general reliability level aimed for was high.

The effects of a new optimum reliability level on load forecasts and outage costs via the price and reliability expectation feedback loops shown in Fig. 3.3 were not investigated; however, the results of the study were found to be relatively insensitive to an arbitrary 10% change in load forecast.

Table 3.1 Outage Costs

Study	Outage costs US$ (1977 prices) (Depending upon the duration of the outage)	
	Residential Consumers	Industrial Consumers
The above (1977)	1.3 to 2.0	1 to 7
Sweden (1948)	0.4 to 0.7	1 to 2
Sweden (1969)	0.7 to 1.5	0.1 to 3
Chile (1973)	–	0.2 to 8
England (1975)	0.5 to 1.5	–
Canada (1976)	–	1 to 9

Table 3.1 gives some comparative values of outage costs obtained from the above study and other studies.

3.4 APPENDIX: SYSTEM COSTS (SC) AND OUTAGE COSTS (OC): A FORMAL DERIVATION

The optimum reliability rule of Chapter 2 is more difficult to apply than the optimum investment and the optimum pricing rules. A framework for using the optimum reliability rule needs to be developed by power-system planners. To help them do this, one way of deriving formulas describing the rule is given below.

3.4.1 Keeping the price constant

Let us consider a power system in which the growth of the demand for electricity taken over a long term period of $(\tau+1)$ years is represented by D_t, where t varies from 0 to τ.
D is a function of other variables; i.e.

$$D_t \text{ is a function of } (P_t, Y_t, R_t^*, \underline{Z}_t) \qquad (3.1)$$

In this expression:

P = electricity price
Y = income variable, i.e. it is a consumption variable corresponding to the level of national economic activity
R^* = level of reliability expected by consumers
$\underline{Z}$ = complex vector of other independent variables affecting the demand for electricity.

Let us assume initially that the values of P_t, Y_t and $\underline{Z}_t$ are fixed exogeneously. This enables us to temporarily decouple the joint optimum pricing and optimum reliability rules of Chapter 2. Later we will relax this assumption somewhat to get nearer to practical conditions.

We can now represent the outage costs OC_t associated with each power system development plan, incurred by electricity consumers and national economy due to outages of power supplies as a function of only the reliability level R_t, the forecast demand D_t, and the reliability expectation R_t^*; i.e.

$$OC_t \text{ is a function of only } (R_t, D_t, R_t^*) \qquad (3.2)$$

OC_t is a 'demand' side cost. If the corresponding 'supply' side cost to meet the electricity demand growth is SC_t, then:

$$SC_t \text{ can be written as a function of } (R_1, \ldots, R_\tau; D_1, \ldots, D_\tau) \quad (3.3)$$

The supply-side costs in eqn. 3.3 correspond to the least-cost solutions for meeting the system load growth. This means that SC_t includes both capital and running costs, plus the cost of kW and the kWh losses in the system, valued at the 'optimum' cost of supply, net of the marginal supply costs of power not able to be delivered due to outages. Since our reliability expectation R_t^* depends on our past experience of reliability over the period (t-s)

$$R_t^* \text{ is a function of } (R_{t-1}, \cdots, R_{t-s}) \tag{3.4}$$

We can now write a simple expression for the net benefit of electricity consumption to consumers and the national economy. We can write the expression in the convenient form of a total present value of net benefit (NB) in discounted cash flow terms:

$$NB = \text{Total Present Value} \sum_{t=0}^{T} \frac{(TB_t - OC_t - SC_t)}{(1+r)^t} \tag{3.5}$$

where TB_t, a function of D_t, is the total benefit of electricity consumption in the absence of outages, and r is the appropriate discount rate. Because of the long economic life of some projects we can make the discounting time horizon T greater than the planning time horizon τ to eliminate terminal values at the end of the discounting period. Otherwise terminal values must be placed upon all cost streams. The question of terminal values is covered more fully in Chapter 8.

In order to identify the marginal conditions which maximise NB in eqn. 3.5 we can formulate a Lagrangian:

$$L = \sum_{t=0}^{T} \left\{ \frac{(TB_t - OC_t - SC_t)}{(1+r)^t} - \lambda_t \left[R_t^* - R_t^*(R_{t-1}, \cdots, R_{t-s}) \right] \right.$$
$$\left. - \mu_t \left[D_t - D_t(R_t^*) \right] \right\}.$$

The three first-order conditions of relevance are then:

$$(\partial L / \partial R_j) = -\left[\frac{(\partial OC_j / \partial R_j)}{(1+r)^j} + \sum_{t=0}^{T} \frac{(\partial SC_t / \partial R_j)}{(1+r)^t} \right] + \sum_{t=0}^{T} \lambda_t (\partial R_t^* / \partial R_j) = 0$$

$$(\partial L / \partial R_j^*) = -\frac{[(\partial OC_j / \partial R_j^*)]}{(1+r)^j} + \mu_j (\partial D_j / \partial R_j^*) - \lambda_j = 0$$

$$(\partial L / \partial D_j) = \frac{[(\partial TB_j / \partial D_j) - (\partial OC_j / \partial D_j)]}{(1+r)^j} - \sum_{t=0}^{T} \frac{(\partial SC_t / \partial D_j)}{(1+r)^t} - \mu_j = 0$$

for j = 0, 1, 2, ..., T.

Combining the three equations immediately above yields:

$$\left\{ \frac{-(\partial OC_j / \partial R_j)}{(1+r)^j} - \sum_{t=0}^{T} \frac{(\partial SC_t / \partial R_j)}{(1+r)^t} \right\} +$$

$$\sum_{t=0}^{T} (\partial R_t^* / \partial R_j) \left\{ \frac{[-(\partial OC_t / \partial R_t^*) + (\partial D_t / \partial R_t^*)\{(\partial TB_t / \partial D_t) - (\partial OC_t / \partial D_t)\}]}{(1+r)^t} \right. -$$

$$\left. \sum_{u=0}^{T} (\partial SC_u / \partial D_t)(1+r)^u \right\} = 0$$

The first term $\{\ldots\}$ of the above expression captures the direct impact of reliability changes on outage costs OC and system costs SC, while the rest of the equation represents the corresponding indirect effects via the reliability expectation R^*. To be practical we must simplify the expression in some way. For example, we might assume to a first order of approximation that $(\partial R_t^*/\partial R_j).(\partial OC_t/\partial R_t^*)$ and $(\partial R_t^*/\partial R_j).(\partial D_t/\partial R_t^*)$ are negligible. In other words we might say that the effect of changes in the reliability level R on both the outage costs OC and the demand D, via the expected level of reliability R^*, are small. Most practitioners would agree that this is not an unreasonable assumption if the range of R is not wide, which is normally true. Therefore:

$$\frac{-(\partial OC_j / \partial R_j)}{(1+r)^j} - \sum_{t=0}^{T} \frac{(\partial SC_t / \partial R_j)}{(1+r)^t} = 0 \qquad (3.6)$$

Eqn. 3.6 indicates that the net benefits to society will be maximised at the point where the marginal outage costs associated with a change in reliability are exactly offset by the corresponding change in supply costs.

We have a formal proof of what we argued out by logic in the main body of this Chapter.

Although this equation relates to reliability in time period j, changes in reliability are likely to be implemented over several time periods. Therefore, a more useful form in which to write eqn. 3.6 would be:

$$\sum_{j=0}^{T} \frac{-(\partial OC_j / \partial R_j).\Delta R_j}{(1+r)^j} - \sum_{t=0}^{T} \frac{(\partial SC_t / \partial R_j).\Delta R_j}{(1+r)^t} = 0 \qquad (3.7)$$

where ΔR_j is a small change in reliability during period j.

Now let us consider alternative long-term power-sector development programmes, where the reliability level of the ith programme in year t is R_t^i. Suppose that power-system development plan $(i+1)$ is derived from plan i, by a small change in reliability given by:

$$R_t^i = R_t^{i+1} - R_t^i \quad \text{for} \quad i = 0, 1, \ldots, n.$$

Examining eqns. 3.4 and 3.7 the corresponding change in net benefits (NB) may be written:

$$\Delta NB^i = NB^{i+1} - NB^i = -\Delta OC^i - \Delta SC^i \tag{3.8}$$

where

$$\Delta OC^i = \sum_{j=0}^{T} \frac{(\partial OC_j^i / \partial R_j^i) \cdot \Delta R_j^i}{(1+r)^j}$$

and

$$\Delta SC^i = \sum_{j=0}^{T} \sum_{t=0}^{T} \frac{(\partial SC_t^i / \partial R_j^i) \cdot \Delta R_j^i}{(1+r)^t}$$

In eqn. 3.8 let us assume that the change from power-system development plan i to plan (i+1) involves an overall unambiguous improvement in reliability; i.e. each component ΔR_t^i is non-negative. In general, this implies correspondingly that $\Delta OC^i < 0$ and $\Delta SC^i > 0$. In this case, eqn. 3.8 yields a simple expression, which is rather important:

$$|\Delta NB^i| = |\Delta OC^i| + |\Delta SC^i|$$

Thus, in order to maximise the net economic benefits of supplying electricity to society in general, but mainly measured from the standpoint of the national economy, the reliability should be increased in successive power-sector development plans, each plan being the least-cost solution, as long as the corresponding decrease in incremental outage costs ΔOC exceeds the increase in incremental supply costs ΔSC. Also, since in our original assumption the total benefit TB is assumed to be independent of the reliability R, the net benefit NB is maximised when the present discounted value of the sum of outage costs and supply costs is minimised. Once again this is a formal proof of the point made previously (see p.49).

Now, let us define the scalar reliability index R^i which characterises the ith system development plan, as follows:

$$R^i = 1 - \frac{\left[\sum_t OE_t^i / (1+r)^t\right]}{\left[\sum_t TE_t / (1+r)^t\right]}$$

where OE is the electrical energy in kWh not supplied because of outages and TE is the total electrical energy in kWh that would have been supplied in the absence of outages. Fig. 3.1 (see Section 3.1.2) shows a typical graph of outage costs OC and supply costs SC associated with different power-system development programmes and reliability levels. The total costs TC = SC + OC is also plotted. As R increases SC rises more and more rapidly; it can be seen that a perfectly reliable system ($R = 1.0$) is not attainable. Correspondingly OC falls towards zero as R increases towards unity. The optimal reliability level R^m is at the minimum point of the TC curve, when the slope of SC is equal to minus the slope of OC, corresponding to eqn. 3.7.

In the above we defined the reliability measure in a very generalised way. Therefore, the selection of the optimum power-system development programme, and the reliability level associated with it in the least-cost solution, is made on the basis of economic cost-benefit analysis. It is worth noticing that it is quite independent of the <u>actual</u> index of reliability. Having said this, it is important to develop reliability indices (see Section 3.2.5) which not only characterise future system performance in a satis-factory manner, but are also:

• meaningful to consumers; and
• could be easily used to determine the costs of shortages incurred by consumers.

From this point of view the so-called 'load-point' indices showing the frequency and duration of outages, voltage reductions etc. referred to individual consumers on a disaggregate basis are the most convenient measures to use (see FAD indices in Section 3.2.5). It is also very important to know as precisely as possible when outages occurred. Since outage costs are generally a nonlinear function of outage duration, we should ideally compute the probability distribu-tion of outage duration. However, a knowledge of the mean duration of outages at specific times is usually sufficient; e.g. the mean duration at the peak periods, the 'shoulder' of the peak and the off-peak periods.

3.4.2 Allowing the price to vary

Let us relax our earlier assumption on the original fixed price of electricity P_t. We assume that the stream of supply costs SC_t^m, associated with the optimum power sector development plan i = m on the first round of working, require us to make quite significant changes regarding the evolution of prices P_t, these themselves being used to determine the initial load forecast. The investment/pricing rules, or the financial requirement of making an adequate annual financial return on fixed assets, may well require us to make previously unforeseen changes in future electricity tariffs in order to compensate for new supply costs. Changes in prices directly affect load forecasts; also the new target reliability levels implied by the first-round optimum power-system development plan i = m may themselves affect reliability expectations (see eqn. 3.1 and 3.2). Hence we must consider the impact of new sets of price and expected reliability levels on the load forecast and the outage cost estimates, when iterating through our above-described model for the second and subsequent rounds. Usually, this procedure will change the whole

total cost TC curve (see the broken line in Fig. 3.1), leading to
a new optimum development plan m' at reliability $R^{m'}$.   Thus we
iteratively consider the direct and indirect feedback effects of
reliability on demand until a more-or-less self-consistent set of
price, demand and reliability levels are determined.

REFERENCES

[1] MUNASINGHE, M.: 'Power system economics, pricing and
     reliability standards; Part II: Optimal reliability and
     system planning'. Paper presented at the International
     Symposium on Electricity Economics and Load Management,
     Imperial College, London, 24-28 March 1980; and WEBB, M.:
     'Power system economics, reliability standards and
     pricing'. From the same Symposium. These papers provide
     the basic framework for this chapter, and especially
     Appendix 3.4; by permission of Imperial College and Dr.
     Munasinghe

[2] MUNASINGHE, M.: Reference 1, p. 1

[3] BERRIE, T.W.: 'The economics of system planning in bulk
     electricity supply', in TURVEY, R. (Ed.): 'Public enter-
     prise' (Penguin Economics Classics, 1968)

[4] Ibid.

[5] CASH, P., and SCOTT, N.: 'Security of supply in planning
     and operation of European power systems'. IEEE Trans.,
     Jan. 1969, PAS. In many ways this is still the most
     comprehensive paper on how plant margins are determined
     empirically in practice

[6] BERRIE, T.W.: 'Margins, risks and costs', Elect. Rev., 15
     Sept. 1967, Fig. 4 and describing text

[7] MUNASINGHE, M., and GELLERSON, M.: 'Economic criteria for
     optimising power system reliability levels', Bell J.
     Economics, Spring 1979

[8] WEBB, M.: 'The determination of reserve generating
     capacity criteria in electricity supply systems'. Appl.
     Economics, 1977, 9, pp. 19-31. The example is taken from
     Webb in Reference 1

[9] GITTINGER, J.P.: 'Compounding and discounting tables for
     project evaluation' (Johns Hopkins University Press,
     Baltimore and London, 1973)

[10] DREYFUS, H.B., BERRIE, T.W., and KNIGHT, U.G.: 'Primary
     system planning in England and Wales for security of
     supply'. CIGRE Proceedings, 21969, 1968

[11] BILLINGTON, R., RINGLEE, R.J., and WOOD, A.J.: 'Power
     system reliability calculations' (Cambridge, Mass. MIT

Press, 1973); HAJDU, L.P., and PODMORE, R.: 'Security enhancement for power systems', Bell J. Economics, Autumn 1979, 11, pp. 177-195. For a detailed definition on component failure see IEEE Standard No. 346-1973, New York, IEEE, 1973

[12] SINGH, C., and BILLINGTON, R.: 'Frequency and duration concepts in system reliability evaluation', IEEE Trans., 1975, R-24, pp. 31-36

[13] BILLINGTON, R.: 'Power system reliability evaluation', (Gordon and Breach, New York, 1970)

[14] NOFERI, P.L., PARIS, L., and SALVADERI, L.: 'Monte Carlo methods for power system reliability evaluation in transmission and generation planning'. Proceedings of the Annual Reliability and Maintainability Symposium, Washington DC, Jan. 1975, pp. 460-469

[15] ALLAN, R.N., and TAKIEDDIN, F.N.: 'Generation modelling in power system reliability evaluation', in 'Reliability of power supply systems', IEE Conference Publ. 148, London UK, Feb. 1977, pp. 47-50

[16] ALLAN, R.N., and TAKIEDDIN, F.N.: 'Network limitations on generating systems reliability evaluation techniques'. Paper A78070-5, Proceedings of the IEEE Power Engineering Society Winter Meeting, New York Jan.-Feb. 1978; MARKO, G.E.: 'Method of combining high speed contingency load flow analysis with stochastic probability methods to calculate a quantitative measure of overall power system reliability'. Paper A78053-1, Proceedings of the IEEE Power Engineering Society Summer Meeting, Los Angeles, July 1978

[17] BROWN, G. Jr., and JOHNSON, M.B.: 'Public utility pricing and output under risk', American Economic Rev., Mar. 1969, pp. 119-128; CREW, M.A., and KLEINDORFER, P.R.: 'Peak load pricing with a diverse technology', Bell J. Economics, Spring 1976, pp. 207-231; SHERMAN, R., and VISSCHER, M.: 'Second best pricing with stochastic demand', American Economic Rev., Mar. 1978, pp. 41-53; and CREW, M.A. and KLEINDORFER, P.R.: 'Reliability and Public Utility Pricing', American Economic Rev., Mar. 1978, pp. 31-40. Crew and Kleindorfer also recognise that there are other costs associated with outages, i.e. the rationing costs of determining how to shed loads at the lowest cost, but these costs cannot be defined well enough to take into account

[18] TELSON, M.L.: 'The economics of alternative levels of reliability for electric generating systems', Bell J. Economics, Fall 1975, 6, pp. 679-694; TURVEY, R., and ANDERSON, D.: 'Electricity economics', (Johns Hopkins University Press, Baltimore, Md., 1977) Chap. 14; MUNASINGHE, M.: 'The costs incurred by residential electrical consumers due to power failures', J. Consumer Res., Mar. 1980; and MUNASINGHE, M., and GELLERSON, M.: 'Economic criteria for optimizing power system reliability levels', Bell J. Economics, Spring 1979

[19] WEBB, M.: 'The determination of reserve generating capacity criteria in electricity supply systems', Appl. Economics, Mar. 1977, pp. 19-31. M. Webb suggests using electricity tariffs as a lower bound on the estimate of outage costs. Clearly that is inappropriate since, even at the margin, the cost of an unexpected interruption in the supply of electricity will exceed what consumers are willing to pay for planned consumption of electricity

[20] TAYLOR, L.: 'The demand for electricity: A survey', Bell J. Economics, Spring 1975, 6, pp. 74-110

[21] MUNASINGHE, M., and SCOTT, W.: 'Long range distribution system planning based on optimum economic reliability levels'. Paper A78576-1, Proceedings IEEE PES. Summer meeting, Los Angeles, July 1978. (Numbers quoted in the text are from Reference 1 by permission of Imperial College)

[22] MUNASINGHE, M.: J. Consumer Res., Mar. 1980, op. cit. in Reference 18

[23] MUNASINGHE, M.: 'The economics of power system reliability and planning'. Published for the World Bank (Johns Hopkins University Press, Baltimore, USA, 1979)

Chapter 4

Load forecasting

4.1 IDENTIFICATION OF IMPORTANT PROBLEMS

4.1.1 Introduction

Within the planning cycle[1] load forecasting is one of the many ways of stating objectives. To put this more precisely, load forecasting can be regarded as answering the question: what amount of electricity should we arrange to supply a specific number and kind of consumer over a specific period of time? Until recently, development plans were rigorously drawn up so as to meet an 'adopted' load forecast[2] at an 'accepted' standard of reliability and a financial rationale for choosing prices.[3] If the investment required cannot be raised, or the money for fuel found, or the consumer will not pay the required price, we must go round the planning cycle again less ambitiously, e.g. using a smaller load forecast, or a lower reliability standard, or both.

Any optimal development plan must fit in with the objectives of the economy, the utility and the consumers, the latter showing their acceptance mainly by their willingness to pay. In this respect electricity is like any other type of goods or services.

Therefore a load forecast is in no way exogenously given in the planning exercise; it must be seen within the context of all planning objectives. A simple layout of the planning cycle in the sector is shown diagrammatically in Fig. 4.1.

4.1.2 Reasons for making load forecasts

We can now examine the purposes of load forecasting within the planning cycle. Electricity forecasts are needed for the following types of study:

(i)   Generation development programmes; circle 2 in Fig. 4.1.
(ii)   Network development programmes; circle 2 in Fig. 4.1.
(iii) Tariff making; circle 3 in Fig. 4.1.

However, this allocation between circles is not practically

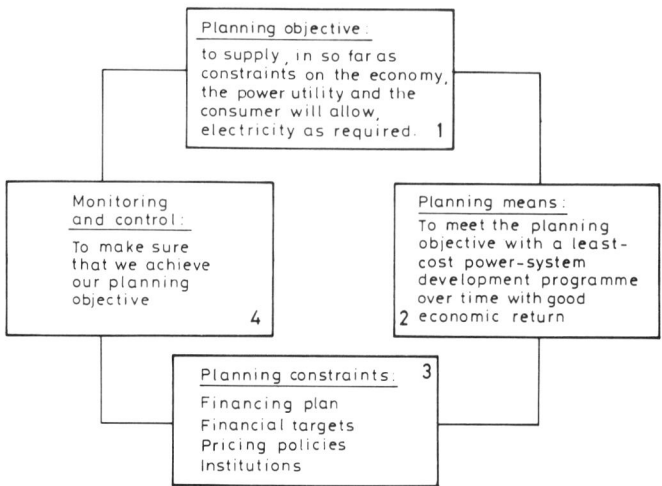

Fig. 4.1   Power-sector planning cycle

possible because we cannot decide anything in the planning cycle without going through all the circles at least once, if only in the 'mind's eye'. A common mistake is not to go round the cycle at least once before finalising a load forecast. This can lead to unrealistic projections.

Most type (i) and (ii) studies use three forecasting parameters: demand, in the power-system sense, kW; duration, again in the power-system sense, kWh; and geographical location. Some type (iii) studies use four parameters: number of consumers, kW, kWh and geographical location. Type (iii) studies use more parameters, including kW, kWh, consumer classes, e.g. domestic, commercial and industrial, and geographical location. A practical problem of load forecasting lies in selecting and researching a balanced number of parameters in the required depth. Using too few parameters, or pursuing them insufficiently, makes the forecast inadequate; too many pursued to too great a depth tends towards unwarranted complication and spurious accuracy.

The power sector needs investments of about £500 to £6000 per consumer, depending upon community sophistication. Power-sector planning mistakes can be costly. Too high forecasts lead to more plant than is required by the Reliability Rule and unnecessary[4] capital expenditure. Electrical plant manufacturers may also be badly affected when plant orders are eventually drastically reduced, as shown by recent experience in the UK. Too low load forecasts prevent optimum economic growth and lead to the installation of many costly and expensive-to-run private generators.

From the above we can appreciate the value of accurate forecasting; above all, we should avoid being persistently wrong in either direction, i.e. too high or too low.

4.1.3 Load forecasting period

The period of the forecast must be appropriate for the purpose of the study. Studies of type (i) require a forecast over a period at least equal to the time taken to bring the development programme to full output, and preferably longer. Thus, for a sector with hydroelectric plants, the forecast period must last at least until all the water available can be used to generate electricity, possibly longer. This brings us to 20 or 30 years after the initial commissioning of some hydro projects, i.e. 30 to 40 years after the forecasts were made. Forecasts for network planning should have a duration at least until most of the network is operating at full capacity. Only 'spot' figures at five yearly intervals are usually necessary for the last 10 - 15 years of any forecast.

Studies (iii) concern tariffs in the immediate future, but these must take into account longer-term effects, especially for marginal cost pricing which is bound up closely with the Investment Rule; they need forecasts for 20 - 30 years ahead.

When looking further ahead than the immediate future we need to handle a volume of numbers which grows in size almost directly proportional to the number of years over which the forecast is made. Many forecasters spend much time getting the forecasts for the early years 'right', but because of the amount of data leave little or no time to deal reasonably well with forecasts for later years. It is a difficult balance; uncertainty in planning increases with the length of time ahead and it could be better to spend more time getting the later years reasonably right, although a forecast far into the future may have little impact on immediate decisions. Time should be allocated to future years only in proportion to their relative impact on immediate decisions; because only decisions taken now are vital, tomorrow's decisions will be taken tomorrow and meantime we hope to have a chance to reassess things. This type of dilemma calls for the use of a mixture of different types of methodology, each balancing their different abilities to cope with future forecasts, as is discussed later on.

4.1.4 Load forecasting principles

Fig. 4.1 showed one view of the 'load forecasting' process turning consumers' hopes for electricity via a planning cycle into a realistic forecast. Most planners agree that forecasting is affected by pricing policy and economic prosperity; economists point out that the willingness and the ability of consumers to pay a given price at any given time in the future are not necessarily the same. Willingness and ability to pay depend on future economic wealth, future income distribution and future prices. The wealth of each person, village, town, country, and indeed the world, are all bound up together in

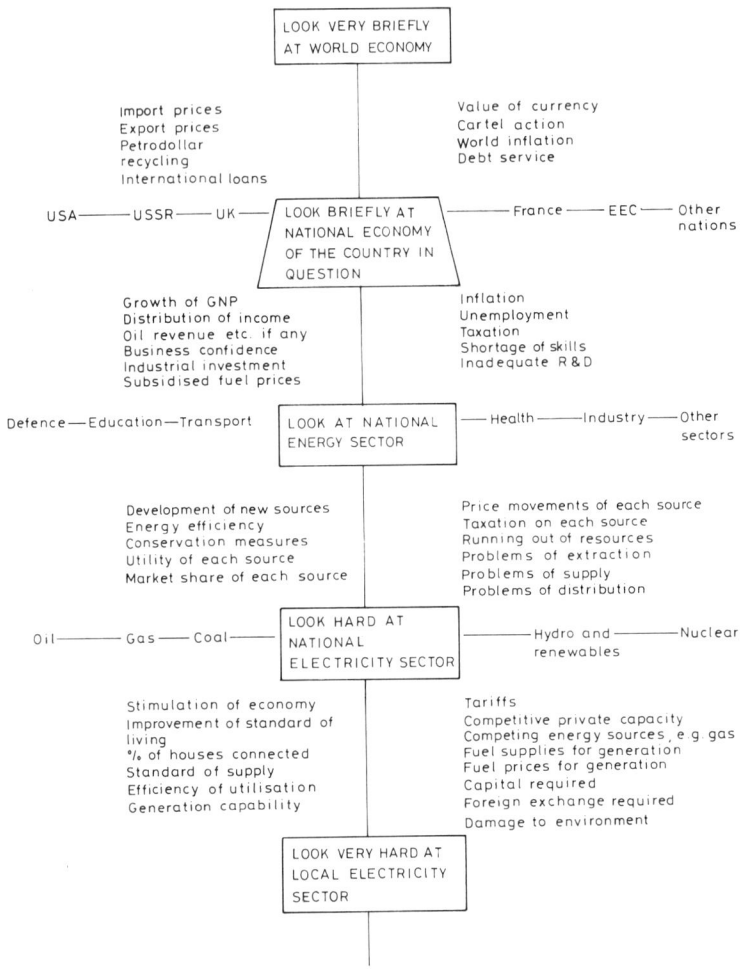

Fig. 4.2   Factors affecting electricity demand

Based on a diagram from BERRIE, T.W.: 'Economic comparison of electricity projects'. Talk given at Seminar at SPRU Unit, Sussex University, 1977

a little understood and highly complex way. Therefore, load forecasting requires a study at all levels of disaggregation, starting very briefly with the world, the national macro-economic and overall energy sector, and finally national and local power sectors. To make proper load forecasts we must study the likely future world-wide and national fuel and electricity prices. In developing countries the likely future connection cost for new consumers is also important, together with prices of fuels which can substitute for electricity and vice versa.

Fig. 4.2 summarises the above. Working downwards we have:

(i)   A very brief look at the world economy
(ii) A brief look at the national economy of the country in question
(iii) A look at the national energy sector in question
(iv) A hard look at the national electricity sector
(v)   A very hard look at the local electricity sector.

Factors affecting the next level down are shown beside the stem of the diagram. However, a change in any factor can have repercussions at all levels, although the weight of impact decreases with the height from the bottom. That is why a very brief look only is needed at level (i), a brief look at level (ii), a longer look at level (iii), and a hard look at levels (iv) and (v).

4.2   LOAD FORECASTING IN PRACTICE

4.2.1   Introduction

A forecaster usually knows in advance when a forecast is required and always has a recent forecast to hand. The good forecaster always watches the underlying forces in the power market and makes a guess at their effect on electricity demand. If a forecast looks wrong after only a few months following fresh information becoming vailable, there may still be time to change some decisions and prevent waste.

4.2.2   Price of flexibility

We must reconcile flexible load forecasting with the multi-parameter forecast of Fig. 4.3. Changing a forecast can require a lot of work; thus we use computer programs to work out and store forecasts. Writing programs and storing information is costly, and there is a balance. Overall, it is more important to be able to quickly modify a forecast than to have an elegant forecasting procedure. Forecasting should be kept simple; several ways of doing this are now given.

4.2.3   Use of statistics

We must always take time to decide which statistics are needed; it is not <u>always</u> necessary to analyse population, numbers of consumers, kWh sales by <u>every</u> consumer class for <u>every</u> year of the

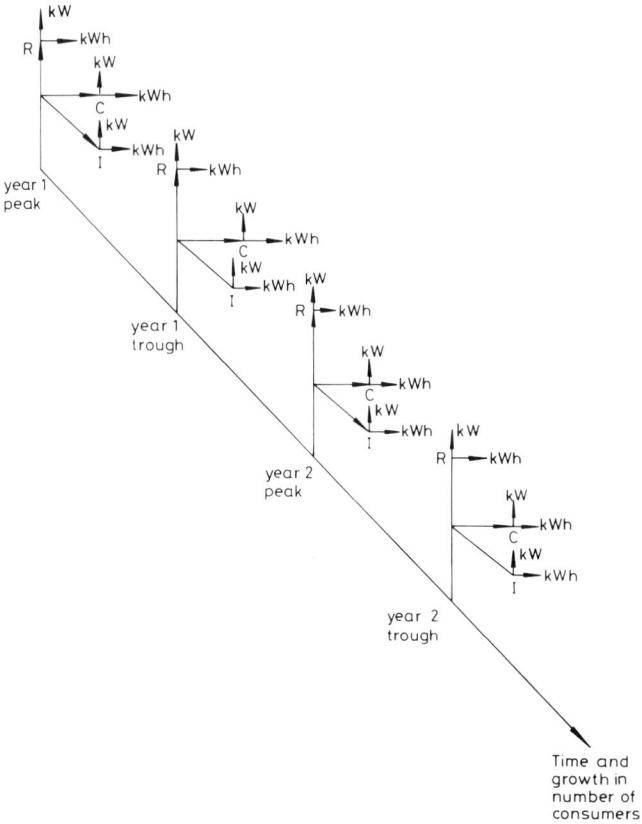

Fig. 4.3   Load forecasting parameter

R: residential consumer
C: commercial consumer
I: industrial consumer

forecast. The forecaster must be especially careful about spurious
accuracy. Implied accuracy is shown in Table 4.1. If numerical values

Table 4.1   Implied accuracy and number of significant figures

Number of significant figures	Smallest number possible	Range of values implied by that number		Maximum error implied
		from	to	
1	1	0.5	1.49	100%-67%
2	10	9.5	10.49	5.2%-4.8%
3	100	99.5	100.49	0.5%-0.5%
	mean number			
1	5	4.5	5.49	11%
2	50	49.5	50.49	1%
3	500	499.5	500.49	0.1%

are expressed to one significant figure and each may on average be
11% inaccurate, then the sum of n numbers having positive or negative
errors each expressed to one significant figure will be 11/n%. This
is 1% if n = 11; expressing a second significant figure on values
no more accurate than say ± 10% can be misleading. Quoting excessive
significant figures also means unnecessary extra work.

Fig. 4.2 indicates that we must look briefly into any national
plans, fuel policy and power-sector reviews. Forecasters should talk
to government and utilities to obtain their current opinions about
the future.

All power utilities have a list of consumers and those waiting
connection or wanting to increase their supply. Forecasters must
talk to current and potential consumers to gauge their prosperity
and their ability and willingness to pay.   They must talk to
householders in towns and villages, shopkeepers, commercial and
industrial managers, and entrepreneurs starting new businesses. To
test for realism they need a second opinion from consultants, bankers
and government ministries.

Forecasters must consult government social ministries, e.g. of
health, for hospitals and water supply; of agriculture, for pumping
loads; of housing, for new estates; of industry for new industrial
loads and estates.

4.2.4   The data base

Historic data forms the bottom layer of the data base; the
forecaster must seek out the past. The latest years of available
statistics are the 'base' years although earlier periods enable

trends to be plotted. We should look as far back in time as the period ahead for which we are forecasting, if data are available.

We must assess the demand suppressed by low voltage or frequency, or a prohibition on electricity use at time of peak demand; suppressed demand shows itself in more private generators, a long list of consumers waiting for connection, or the obvious need to uprate capacity.

The insert below shows the data to be collected; not all is equally available or relevant; a forecasting skill is determining what data is relevant.

DATA TO BE COLLECTED FOR LOAD FORECASTING

1. Gross national product; allied economic indicators
2. Population, split into urban and rural
3. Number of houses, split into urban and rural
4. Commercial activity indicators; various productivity indices
5. Industrial activity indicators by industry, including productivity indices and coefficients
6. Energy and fuels uses and prices
7. Self-generation capacity
8. Suppressed kW and kWh p.a. for each tariff category
9. Electricity tariff for each tariff category
10. Number of consumers of each tariff category, e.g. domestic, commercial, industrial
11. kW of maximum demand on-peak by each tariff category
12. kW of maximum demand off-peak by each tariff category, kWh p.a. by each tariff category
13. Political constraints
14. Institutional constraints
15. Financial constraints, e.g. how much capital is available to implement the load forecast; financial borrowing limits

---

Data collection involves accumulating the 'right' material, then translating this into a load forecast. The credibility of the forecast depends not only on data credibility but also on the credibility of the analysis process. The person who specifies what data to collect must also be reponsible for making the forecast. If two people are involved the reasoning of the first must be transferred as exactly as possible to the second.

4.3   THE THREE BASIC METHODS OF LOAD FORECASTING

4.3.1   Introduction

Many systems have a well established growth rate; e.g. it doubled every ten years in England and Wales at one period. Such systems show an 'equilibrium of growth', and forecasts of kW and kWh for five to ten years ahead are the most important. The system is of sufficient size and integration that large base-load steam and nuclear plants are continually being added and such plants take five to ten years from selection to commissioning.

On small rapidly growing systems the important forecasts are for the next two or three years, for this is the lead time for gas turbines and diesel plant. When hydroelectric stations are involved, forecasts for beyond the tenth year ahead are important.

When determining a development programme, forecasts are needed for more than ten years ahead: about 15 to 20 years for small systems; 25 to 35 years for large systems; and possibly 50 years ahead for systems with appreciable hydroelectric plant.

Using a variety of methods is the principal way of making accurate forecasts,[5] as this permits adequate cross-checking. Basically, three methods are used.[6]

4.3.2   Trend forecasting

This method 'fits' trends to actual kW and kWh by regression analysis. It is easy to be misled in the fit of a trend, because cause and effect are not considered separately. The errors introduced by trending randomly cancel out, the effects of particular errors pass unnoticed, and a new dominant cause wil not be spotted. Again, a number of different algebraic expressions produce trends fitted to past experience, give identical forecasts for a few years ahead, and yet give vastly different forecasts after five years ahead (see Fig. 6.1 of Chapter 6).

The trend method should be developed to detect when new dominant causes are just emerging and to decide if these will mean departing from past trend. Unfortunately no one has been able to do this. Nevertheless trend forecasting is a necessary adjunct to the two other methods, especially as it can make separate forecasts of both kW and kWh; the other two methods derive forecasts of kW from forecasts of kWh.

4.3.3   Market forecasting

The main advantage of a market forecast is that it is made by a commercial engineer who knows the sector at first hand. An undoubted disadvantage is that it has been improved very little over the last 50 years; hence the results must be compared with those from at least one, and preferably both, the other methods.

The market forecasting method means finding out at first hand the likely market for as far ahead as we can see and as precisely as we can. It relies on asking consumers the right questions. The most succint and clear explanation is given in Reference 3 under 'Area Board forecasts'.

Other attractions of the method are:

(i)   It is similar to the method used in other industries and therefore is likely to be understood by businessmen. The customer is approached to find out his present and expected future electricity requirements and what he is prepared to pay.

(ii) The method is employed by those at the 'cutting edge' of sales, who know the sector and market well.

(iii) Some form of sophistication can be added, trending particular customer's requirements, or correlating consumer classes' requirements with growth in incomes and expenditures; i.e. the other two forecasting methods can be 'internalised' by a commercial engineer with a rudimentary knowledge of regression analysis, or econometrics.

The other two methods cannot similarly internalise their counterparts.

The errors from the method come from using it:

(i)   Beyond about three years ahead in areas which have been on supply for some time.

(ii) Beyond about two years ahead for areas newly on supply.

This is because it relies on future plans which may change substantially at short notice. Unfortunately errors introduced by consumers are nearly always in the optimistic direction.

When market surveys are being done independently by a number of interconnected utilities, then commonality of some assumptions is needed:

(i)   Macro-economic advice on national growth rates; growth rates of main sectors; national energy policy; etc. This information comes from government. Thus there is some tie-up between the market-forecast method and the third method, the macro-economic method.

(ii) Rules for combining separately made kW forecasts; there are special difficulties because diversity of incidence in maximum demands between different sections of an interconnected system means it is difficult to calculate a total 'simultaneous' system maximum demand. Most utilities use tables of 'diversity factors' by which to multiply the kW of individual parts of the system in order to arrive at an 'after-diversity maximum demand' for a composite part of the system. These after-diversity maximum demands can then be added to give the simultaneous system maximum demand on that composite part of the system. There is also diversity between various voltage levels, types of subsystems by geography, kinds of consumer, etc. and this is tackled similarly.

Certain improvements in the market-survey method should be made for rural and peri-urban areas:

Firstly, we must learn the threshold per capita income below which electricity will not be purchased. This is part of pricing, including 'life-line' tariffs[7] (see Chapter 9).

Secondly, we must become expert in consumer preference, the elasticities of demand with respect to price and the cross-elasticities of demands between various fuels with respect to the relative prices of the fuels. Little is known about these elasticities, especially the cross-elasticities; they greatly influence lower income groups in peri-urban and rural areas. Better knowledge of the threshold income gives a more accurate forecast of the number of new consumers likely to be able to afford connection as income expands or is redistributed. A better knowledge of the elasticities also leads to a better forecast of the rate of growth of electricity per consumer connected, and a better knowledge of both these factors helps to extend the time horizon for the market forecast beyond the very short term of two or three years.

Today's use of marginal-cost pricing makes it difficult to use money prices for fuels in forecasting. Over the long term the money prices of all fuels will be brought up to approximately the level of their marginal costs of supply.[8] Many countries have declared this to be part of national energy policy. However, the practice at present of basing a forecast on money prices for fuels leads to many forecasting distortions, e.g.:

(i)   Output from privately owned generators, attractive on present-day prices, may not be so from a marginal-cost point of view.
(ii) The substitution of other fuels by electricity may look attractive in money prices but may not be good national fuel policy from a marginal-cost point of view.
(iii) Electrical motor drives, attractive from a money-price viewpoint should often not replace mechanical drives from a marginal-cost viewpoint
(iv) In rural and peri-urban areas the substitution of electricity for kerosene lighting, heating etc. may not be economic in present money prices in the long run.

There are many time-lagged variables in forecasting. This is one reason why market forecasts prove optimistic. There is a natural rhythm to development[9] despite political, social, economic and pricing pressures. From the data available from long-established utilities it should be possible to determine the numerical value of time lags and the time-lagged variables. This would considerably improve the market forcasting method, and help to extend its horizon beyond the three years mentioned earlier.

Finally, to make market forecasts which give decision makers

confidence we must credibly integrate the forecasts for established
and 'greenfield' areas, to arrive at a total forecast.[10]

4.3.4   Macro-economics and econometric forecasts

The use of macro-economic and econometric forecasts in the power
sector is not new.[11]   Utilities may not be involved in such
forecasting; others may be asked to work directly with government
to make econometric forecasts for the energy or electricity sectors.
Some macro-economic data must be used by both the trend and market
forecasters. However, econometric forecasting goes much further by
producing an independent forecast based directly on relationships
between the following items:

(i)   Growths in GDP and its components, disaggregated by sector,
        e.g. the energy sector and the power sector
(ii)   Growths in overall industrial production by type of industry
(iii)   Growths in overall commercial activity, e.g. shops, offices,
        etc.
(iv)   Growths in overall population, income, expenditure, by elec-
        tricity consumer class, etc.
(v)   Growths in the sale and utilisation of domestic appliances.

Whilst combining cross-sectional analysis between the different
sectors with trending, such econometric forecasts mainly seek
correlations between the growths in the national economy, the total
energy sector and the power sector. Its full use as a forecasting
method has been delayed because it has been mainly used by economists
whereas the other two methods are used by engineers, and the latter
are usually in charge of planning.
Econometric forecasts depend upon having valid input-output tables
for the economy, and knowing key econometric ratios.

4.3.5   Uncertainty

The importance of uncertainty in load forecasting was brought out
in the comprehensive study done by the World Bank in 1969/70 (see
Reference 5). Before that, tables like Table 4.2 were published.
The World Bank study estimates: (i) the 'random' i.e. unforecastable,
and (ii) the 'systematic', i.e. more forecastable, components of load
forecasting errors; it suggests possible steps to improve accuracy.
Although done in 1969 it is still worth perusing. The study shows
that percentage errors in forecasting have a 'normal' distribution
around an overall mean of zero, and therefore are just as likely to
be too high as too low. Such forecasts have a standard error as high
as 38%; this may be because forecasting is biased or simply because
errors are large and, individually, highly random.

Table 4.2 Errors in forward estimates of maximum system demand
(in standard weather conditions) in England and Wales
expressed as a percentage of the estimates*

	Number of winters ahead to which estimates relate							
Year of estimate	1	2	3	4	5	6	7	8
1947	-3.4	-3.3	-7.1	-8.8	-5.4	-6.0	-6.4	-8.5
1948	-3.6	-8.0	-9.4	-6.3	-7.3	-8.0	-10.3	-14.5
1949	-6.6	-8.6	-4.8	-4.9	-4.9	-7.5	-11.0	-11.9
1950	-6.3	-1.9	-1.5	-1.2	-2.7	-5.7	-6.8	-8.7
1951	+2.1	+3.2	+3.0	+1.2	-2.2	-2.6	-3.8	-6.6
1952	0	+1.2	+0.6	-2.8	-2.6	-4.4	-6.6	-9.8
1953	-0.6	-2.4	-4.5	-4.8	-6.5	-8.6	-12.3	-16.4
1954	-1.8	-3.9	-4.2	-5.4	-8.1	-11.3	-15.9	-16.1
1955	-2.0	-1.8	-3.5	-4.8	-7.8	-11.3	-14.9	-17.7
1956	+0.8	-0.1	-1.1	-4.2	-7.3	-10.7	-13.3	-12.2
1957	-0.4	-1.0	-3.8	-6.8	-9.9	-12.4	-11.4	-11.2
1958	-1.0	-3.2	-6.2	-8.9	-11.3	-10.2	-9.8	-10.5
1959	-2.4	-5.0	-7.2	-8.8	-6.9	-5.6	-5.1	-1.6
1960	-1.6	-4.2	-6.1	-4.6	-4.0	-3.6	0	
1961	-0.7	-1.4	+0.8	+2.6	+3.4	+7.1		
1962	+0.7	+3.6	+5.8	+7.3	+11.5			
1963	+4.2	+7.8	+10.4	+15.6				
1964	+3.7	+6.8	+12.4					
1965	+1.0	+5.4						
1966	+5.0							
Mean error	-0.6	-0.9	-1.5	-2.7	-4.5	-6.7	-9.1	-11.2
Standard deviation ,%	3.1	4.7	6.1	6.7	7.2	8.5	10.5	12.5

+ indicates overestimate   - indicates underestimate

The errors in some cases were indeed large. Over 60% were in error
by more than ±20% and over 30% by ±40%. The same conclusions applied:
(i) for forecasts of kW or kWh; (ii) regardless of the forecast
interval; (iii) in forecasts for both large and small systems; and
(iv) for systems in all growth rates from 5% to 20% p.a.
In the very short term, i.e. one or two years ahead, the 'random'

*Source: BERRIE, T.W.: 'Margins, risks and costs'. Elect. Rev.,
15 Sept. 1967, Table 1 Reproduced by permission of the Electrical
Review

errors, or those which <u>cannot</u> be reduced, were very large at about 50% per year. However, what seems vital is to reduce 'systematic' errors, i.e. errors which <u>can</u> be reduced, for these were shown to reach about 50% on average, after two years, and about 60% to 85% after five years. The study indicates that if a forecast is high (or low) in the short term, it is just as likely to be high (or low) in the long term, although it may be easier to reduce long-term errors because more time is available.

New forecasters are likely to make the worst forecasts and developing countries do make worse forecasts than developed countries. Although forecasting errors in developing countries shown in the study are about twice those in developed countries, the reasons for this are not simple. Forecasting accuracy does not appear to be improving; forecasts made in the 1950's seem no less accurate than those made in the 1960's or early 1970's.

To cope with uncertainty some form of sensitivity[12] or probability analysis[13] is needed for the main parameters: load factor, deviations from past trends, and national planning factors.

4.3.6 Economic load forecasts: Cost-benefit criteria

Within the planning cycle forecasts will always theoretically include <u>only</u> those loads economically justifiable. In poor countries and in rural areas, where initially loads are small and supply costs large, one must check this criterion at the beginning and not at the end of the forecasting process, when checks are normally made. This means working out the long-run marginal cost of supply and the economic worth of the load simultaneously. Such figures should be worked out for every individual load; because of the 'lumpiness' of installing equipment and connecting groups of consumers a good deal of averaging must be done. The basic approach is given below.

RULES FOR LOAD FORECASTING IN RURAL AREAS

(i)   Look briefly at the growth potential, population and other macro-economic factors.

(ii)   Produce a rough, first design of the network to fit (i).

(iii) Work out typical cost-to-benefit ratios, using the opportunity cost of capital as the discount rate, to determine some 'ranking' of consumers or areas by overall economic worth, and those areas or consumers whose cost-to-benefit ratio is below unity. Consideration of other constraints, e.g. total investment available, selection of special consumers, is done at this point. Revenues can be taken as a (low) first measure of benefits.

(iv)   Refine the system design in (ii) in the light of (iii) and repeat the procedure by judgment until a forecast is produced which appears likely to be near the economic optimum and fits inside other constraints.

REFERENCES

[1] BAUM, W.C.: 'The project cycle', Finance and Development, Dec. 1978; and Chapter 2 of this book

[2] BERRIE, T.W.: 'The economics of system planning in bulk electricity supply, margins, risks and costs', Elect. Rev., 15 Sept. 1967

[3] EDWARDS, R.S., and CLARK, D.: 'Planning for expansion in electricity supply'. Proceedings of 14th British Electrical Power Convention, 1960

[4] First Report from the Select Committee on Energy, Session 1980-81. Feb. 1981, HC 114-I, H.M. Stationery Office, London, UK; also Report on Central Electricity Generating Board, by Monopolies Commission, May 1981, HC 315, H.M. Stationery Office

[5] Some very interesting facets of load forecasting in the power sector are given in 'Ex-post evalutation of electricity demand forecasts'. World Bank Economics Department Working Paper 79, June 1970 (by permission of World Bank)

[6] BERRIE, T.W.: 'The economics of system planning in bulk electricity supply', in TURVEY, R.(Ed.): 'Public enterprise' (Penguin Modern Economics, 1968) pp. 173-178; but see a somewhat different grouping in RHYS, J.M.W.: 'Demand forecasting'. Paper presented at International Symposium on Electricity Economics and Load Management, Imperial College, London, UK, 24-28 March 1980

[7] The so-called 'life line' rates for goods and services are those tariff levels which can be paid by even the poorest if they need the goods and services to support life. World Bank Annual Report, 1979

[8] TURVEY, R., and ANDERSON, D.: 'Electricity economics'. Published for the World Bank (Johns Hopkins University Press, Baltimore, USA, 1977)

[9] 'North south - A programme for survival'. Report of the Brandt Commission, 1980

[10] 'Rural electrification'. World Bank Sector Working Paper, 1966

[11] See, for example, the wording of some Electricity Acts in both developed and developing countries, which state that due account must be taken of economic prosperity when formulating plans for the power sector

[12] POULIQUEN, L.: 'Sensitivity analysis of projects'. World Bank Occasional Paper 10, 1969

[13] REUTLINGER, S.: 'Techniques for project appraisal under uncertainty'. World Bank Occasional Paper 11, 1970

Chapter 5

Fuels and generating plant mix

5.1 FUELS AND GENERATING PLANT

5.1.1 Introduction

The fuels used for generation[1] may be determined more by indigenous fuels than by the availability and prices of world fuels. Countries which had invested heavily in their coal industry went on using coal for power stations even when oil prices were very low. Since the oil price rises, coal usage has been further encouraged and several new coal deposits have been developed, not only in countries where coal has been extensively used, but also in countries where coal resources have hardly been tapped. As we saw in Chapter 1, a world coal trade might be developing. If this is so there will soon be a world market price for coal, or at least regional market coal prices, which we can use in generation planning.

When calculating the optimum mix of generating plant we must use economic costs rather than money costs. These economic prices sometimes involve 'shadow pricing' because they must reflect the cost to the economy; money costs often contain subsidies and taxes, which are but internal transfers in the economy.

Bearing in mind the fuel available (water for hydro plant), the size and kind of generation on offer and the shape of the annual load curve, the optimum plant mix can be established for some year(s) ahead using one of the models described in Chapter 7. It is convenient to start with a year about 10 to 15 years ahead because, especially on rapidly expanding systems, the power system can be regarded in that year as in a 'green field' situation, i.e. we can ignore today's plant. A simple comparison of the total, i.e. capital plus running (including fuel) costs, of each type of generation over a range of load factors enables us to find operating regimes for which various types of plant have minimal total cost. On a large thermal system the result would be something like that shown in the insert (see also Fig. 7.5).

NATURAL LOAD FACTORS ON A LARGE THERMAL POWER SYSTEM

Plant type	Has minimum total cost over range of load factor
Gas turbines	0 - 10%
Diesels	0 - 15%
Pumped storage	10 - 15%
Combined cycle	15 - 40%
Conventional steam	50 - 75%
Nuclear	70 - 85%

Capital costs including interest during construction are usually transformed into specific (per kW) annuitised annual charges, using an interest rate equal to the discount rate and an annuity life equal to the economic life of each plant. To these annual charges we add the annual specific (per kWh) running costs. The total costs of the various plants per kWh output at a particular load factor are then compared.

It is more difficult to use the above method for a mixed hydro-thermal system; we need to have some empirical rules for times when the hydro plant will be used. This method is no use at all for a system which is mostly hydroelectric. Determining when hydro-electric energy will be most effective on a mixed hydro-thermal system can be quite difficult. We must know at least in outline:

(i)   The total annual (a) average, (b) firm, (c) minimum, and (d) maximum kWh likely to be available from each hydro plant.

(ii) The periods of the annual hydro plant-duration curve when the above kWh can be fitted in to the system annual load-duration curve to maximise the savings on otherwise operating the most expensive thermal plant.

To do this we must find some means of fitting a plant-duration curve into a system load-duration curve; the two curves are different because of (i) losses in system equipment, (ii) the plant margin to cater for outages and errors in load forecasting.

Data showing the annual usage of a generating plant, or of the total generating plant against time, are usually expressed in the form of a plant 'duration' curve showing the number of occasions when the output from that plant is equal to or above a particular level, and we may use an 'integrated' plant duration curve. Appendix 5.3 deals with plant and load-duration curves.

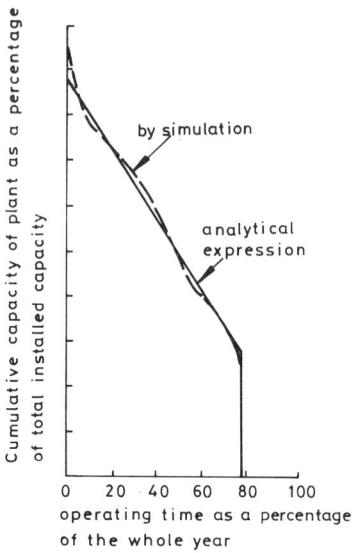

Fig. 5.1   Plant duration curve

BERRIE, T.W.: 'Development of generating plant mix', Elect. Rev., 22 Sept. 1967. Reproduced by permission of Electrical Review.

An important feature of all the curves is their non-linearity as shown in Fig. 5.1, which also shows that the curve can often be made linear by using an analytical expression.   Table 5.1 shows the same information in tabular form.

Once, in the manner described above, we have found the optimum generation plant-mix assuming a 'green field' situation for (say) 10, 15 and 20 years ahead, we must find the optimum way of proceeding from the existing generation to these optimum future plant mix(es). Compromises will need to be made and the plant mix will never be quite optimal; it may at some points be far from optimal, e.g. if load forecasts or fuel costs are seriously in error.   The details of how we get from the existing plant mix to the future optimal plant mixes, usually by means of a computer model, are given in Chapter 7.

5.1.2 Background plan

The above process seldom enables the economic merits of individual generating plants to be assessed; it only gives us a 'background plan', but we need such a plan to meaningfully choose between individual generating stations using 'marginal analysis' as

Table 5.1 Plant-duration curve

MW	h	MW	h	MW	h	MW	h	MW	h
3000	1	2700	12	2400	24	2000	360	1400	192
2975	1	2675	12	2375	48	1950	456	1350	168
2950	1	2650	24	2350	72	1900	696	1300	144
2925	2	2625	12	2325	48	1850	768	1250	108
2900	2	2600	24	2300	48	1800	912	1200	88
2875	4	2575	24	2275	48	1750	1032	1150	101
2850	3	2550	24	2250	72	1700	468	1100	144
2825	4	2525	36	2225	96	1650	372	1050	108
2800	4	2500	36	2200	96	1600	276	1000	48
2775	8	2475	24	2150	168	1550	264		
2750	8	2450	48	2100	192	1500	288		
2725	12	2425	48	2050	264	1450	216		

Peak : 3000 MW
Reserve as a percentage of peak: 20%
Annual rate of increase of load: 8%
Annual discounting rate: 10%
Period of study: 10 intervals of length 2 years
Post-horizon period (in which load is considered to be constant):
20 years

Source: LAUGHTON, M.A.: 'Electrical power systems - Models for expansion planning'. Paper given at International Symposium on Electricity Economics and Load Management, Imperial College, London, 24-28 March 1980. Reproduced by permission of Imperial College, Power Systems Group

described in Chapter 7. Plant mixes are not concerned with transmission and distribution.

There are problems associated with new types of generation, e.g. combined-cycle plant and gas turbines burning heavy fuel oil. With new plant types there are 'running in' costs and initially higher capital costs and lower availabilities.[2] We must specially allow for fuel economies made by new plant types and make full studies of the economics of new nuclear fuel systems.

Coupled with the plant-mix problem there often occurs the fuel transportation problem, i.e. the geographical relationship between power stations and fuel supplies. For coal we set up a matrix of collieries supplying coal to power stations. By using linear programming we establish the optimum transport paths between collieries and power stations.[3] For nuclear power stations the fuel-supply cycle requires detailed optimisation of processing, transport, storage and waste disposal.[4]

Day-to-day power-system operation minimises the total system fuel plus other operating costs.[5]   This does not necessarily either minimise total fuel usage or maximise overall fuel efficiency to the economy. For example, coal is likely to remain cheaper than oil in money terms, and many low-efficiency coal-fired generating stations will thus be operated in preference to even the most efficient oil-fired stations. However, this may not mean optimal fuel usage to the economy because of environmental and conservation factors. Such effects are overcome when shadow pricing of fuels is done for system operation. Nowadays most governments are attempting to make all internal fuel prices reflect world prices, and the need for shadow pricing is diminishing.

5.1.3   Autogeneration and cogeneration

Chapter 9 describes the tendency today to consider the joint planning of all generating stations connected to the public supply, regardless of ownership. In the past we have considered extending our power system mainly by large new power stations all owned by the power utility. With the high price of fuels we must today consider separately owned autogeneration, and combined heat and power co-generation which may be environmentally less damaging and, built in conjunction with other heat uses, thermally much more efficient. The cogeneration plant at Hereford in the UK uses diesel engines with waste-heat boilers to supply both electricity and factory process-heat requirements. There are many other 'combined' power uses, e.g. organic wastes to produce methane to produce electricity from internal combustion engines. Major institutional and managerial problems are usually encountered with such 'combined' generation. A more widespread adoption of the inter-active load control described in Chapter 9 will enable auto-producers and cogenerators to play their proper rôle on the system.

5.2   DETERMINING BACKGROUND PLANS

5.2.1   Plant to load balances

Section 5.1 described the finding of the optimum plant mix(es) at some date(s) in the future. Generating-plant to load balances at an economic reliability standard must then be constructed for the intervening years between the existing situation and that required in the future. At this stage the balances on a large system will be purely numerical, giving little consideration to individual stations; the balances will be adjusted for their sizes later.

On smaller systems some attention must be paid to individual stations even at the first 'trial' plant to load balance because the operation of small systems may be fundamentally affected by the type and size of any generating plant. This means that more than one iteration will usually be needed to form a balance, and there will be some years when there are either excesses or deficits in the balances because of plant 'indivisibilities', i.e. the optimum size of plants being larger or smaller than the size needed to keep the balance for a particular year.

5.2.2 Basic situation

When determining a background plan by fuel and plant size from a 'green field' situation, all equipment being from the same vintage designs, only a quite simple method[6] is needed. The plants will differ in investment to running cost ratios and in their technologies, but not in their degree of obsolescence. Also there are no 'sunk' costs. On the larger systems we take some gas turbines, some diesel and some hydro stations (see Section 5.1.1) for the peak; then we take a large block of coal-fired and nuclear plant to meet the base-load. Finally we fill the middle-load with a mix of hydro, combined-cycle and diesel plant. On the smaller systems the system size will rule out some types of plant altogether because of their optimal size, e.g. steam and nuclear plant.

Having found the optimum plant mix from some years ahead, we then see how to get there from the present plant mix. The approach will take account of five main factors[7]:

(i)   The present plant mix will not be optimal because of unforeseen errors made in the past.
(ii)   Existing plant may be used more economically, rather than buying new plant for a particular system loading, e.g. peak load times.
(iii)   Making investment has short-term effects on the financial cash flows of the utility.
(iv)   Foreign exchange and local costs are needed to implement any development programme.
(v)   The need for a satisfactory economic return on the investment.

Most power-system planners do not go beyond (i) and (ii) although this situation is now changing. We ourselves deal only with (i) and (ii) in this Chapter; (v) is dealt with in Chapter 8, whilst (iii) and (iv) are not part of the subject of this book.

5.2.3 Fuel cost savings

Any method of determining generation development programme uses the concept of 'fuel cost savings'. Those familiar with complex systems of any type, e.g. road, rail and telecommunications, are usually familiar with the concept, and it is part of many business systems, e.g. businesses with a 'chain' of factories or retail outlets. Fuel cost savings in power systems come about as a result of technological advance together with ability to use fuels more efficiently. Generating plant installed today usually has a higher thermal efficiency than existing similar plant. The fuel costs of new plant per kWh generated are therefore lower than those of similar existing plants. This higher efficiency is often obtained at the expense of increased capital cost as we will see in Chapter 8.

As we shall see in more detail in Chapter 8, generating plants are operated in an 'order-of-merit'[8], i.e. an order to be brought onto or taken off the system to meet the growing or declining system load at any time. We always use the generation with the lower running cost first at any point in time.[9] The 'merit-order' is influenced in no way by the capital cost of the plant, which is a 'sunk' cost. The merit-order looks something like this; (first equals best)

First: Nuclear plant
Second: Modern larger coal-fired plant
Third: Modern larger diesel plant, relatively modern and medium-sized coal-fired plant, and combined cycle plant
Fourth: Modern larger gas-turbine plant, older smaller coal-fired plant, and modern larger oil-fired plant
Fifth: Medium and smaller oil-fired plant
Sixth: Smaller gas-turbine plant, e.g. with aero-engines.

Order-of-merit cannot be directly applied to hydro-electric plant; such plant is operated to make maximum savings in system fuel costs.

Order-of-merit means that plant with the lowest fuel costs, nuclear and modern large coal-fired plant, will normally operate at 'base-load', i.e. as continuously as considerations like maintenance will allow. Plant with the highest fuel costs, aero-type gas turbines, will operate as infrequently as possible, i.e. only to meet the peak demand of the year. Thus the significant savings made by a new generating plant is the fuel cost savings made by the integrated change it makes to the fuel costs of the remaining plant in the merit-order. When a new plant is installed it has an effect upon the merit-order to a greater or lesser extent depending upon the difference in running cost per kWh produced between the new plant and this remaining plant. The introduction of aero-engine gas turbines will make little or no change to an existing merit-order. On the other hand, the introduction of a base-load plant will have an effect on the entire merit-order; it displaces the plant which was previously the highest, which therefore runs for less hours than it otherwise would have done. This effect cascades down the merit-order, with all plants below the new base-load plant in the merit-order displacing the plants just below them, which latter plants therefore run somewhat less. The summated result is a saving in total system fuel cost due to the introduction of the new generating plant, i.e. a 'system fuel cost saving'.

For plant designed to run at load factors between base load and peak load, the 'fuel cost savings' made by introducing the new plant will be less than those for a base load plant. The amount that these fuel cost savings are less will depend upon the position of the new plant in the merit-order.

The arithmetic of working out system fuel cost savings made by introducing only one generating plant for one year is quite large; yet assessments of fuel cost savings are required for each year throughout the economic life of any new plant. Also, generating plant introduced <u>after</u> the new plant are likely to be even more efficient than the latter and the fuel cost savings made by a new generating plant will tend to decrease over its life. These factors require the calculations to be done by a digital computer as in Chapter 7. The computer simulates the merit-order loading of all generating plant, old and new, theoretically for every year of the economic life of the new plant, but in practice for 'spot' years only, the fuel cost savings for intermediate years then being found by interpolation.

Annual fuel cost savings are usually present-valued at the test rate of discount[10] over the economic life of the plant. The present-valued annual fuel cost savings are then summed to a total present value, which can then either be directly compared with the capital cost or annuitised over the economic life of the project at an interest rate equal to the discount rate, in order to make comparison with the annual charges. We shall return to this particular subject in Chapter 8.

5.2.4 Approximate optimum plant-mix in the background plan

To determine the optimum plant-mix by size and type of fuel in the background plan we compare annual capital charges to annual running costs for each type and size of generating plant. This is best done by constructing a diagram,[11] which starts with (i) deciding which type of plant is optimum for different load factors of operation as in Section 5.1.1, and then (ii) deciding the optimum proportion of any particular plant type. This gives optimal plant-mixes at particular points in time, e.g. 10, 15, 20 and 25 years ahead. At this stage we may still not have enough information to make a firm decision to go ahead with an individual new generating station; some refinement of the background plan may be necessary.

5.2.5 Marginal analysis

By repeating all or part of the planning cycle, improvements can be made to the background plan. Sensitivity of the plan to different assumptions concerning load forecasts, fuel prices, capital costs, test rate of discount etc. gives an indication of the plan's robustness. Once an acceptable degree of robustness has been achieved it is normal to refine the background plan by making small changes in turn to each new generating plant type. Such refining is the 'marginal analysis' we meet more fully in Chapter 7. Any change which reduces the total present value of the capital plus running system cost over the next (say) 30 years is counted an improvement and accepted. A mathematical model of the power system is usually needed to do the calculations.[12]

Some system planners believe that it is possible to carry on the process of refining the background plan by introducing in turn small amounts of plant closest in characteristics to each individual plant alternative, in order to directly compare these alternatives. The more the total present-value of the system cost is reduced by introducing any generating plant type the 'better' is that alternative. Whether or not alternative individual plants can be validly compared in this way it is the cost framework of the background plan which must be used as a 'benchmark' to judge these alternatives. We make this assumption throughout Chapter 8 which deals directly with marginal analysis and the comparison of individual alternative generating plants.

5.3 APPENDIX: LOAD AND PLANT DURATION CURVES

5.3.1 Introduction

The loading characteristics of power systems is often given as a continuous plot of power (kW) against time extending over a period of, say, a year. This interests both system operators and system planners alike. Both are concerned with:

(i)   evaluating the total requirements of consumers

(ii)   finding the best way of meeting these requirements.

For many purposes a sequence of 365 daily curves is an inconvenient form of presentation. The user will often organise the data in such a way as to clarify particular parts of the demand-against-time curve. In the process other information may be suppressed and there is always some risk involved when using simplified diagrams for purposes other than those for which they were drawn. The possibility of error can be illustrated by considering the procedure of 'stacking' the generating plant in a 'load-duration curve'. Here there are two common sources of error:

(i)   The realities of system operation require a spinning reserve of capacity to cater for forced outages. This is not revealed by the system load-duration curve.

(ii)   The maintenance of plant requires that it should be taken out of service for periods of real sequential time measured in days or weeks. The system load-duration curve is scaled in collated hours, or, strictly, instants in time.

The generator stacking process is a simple one and useful in studying background plans and plant-mix problems. However, its validity depends on either:

(i)   assessing the errors introduced by use of a simple load-
        duration curve (Fig. 5.2 curve 3) and allowing for them; or

(ii)   using a modified curve which compensates for the effects of
        spinning reserve and planned and forced generation plant
        outages (Fig. 5.5 curve 26).

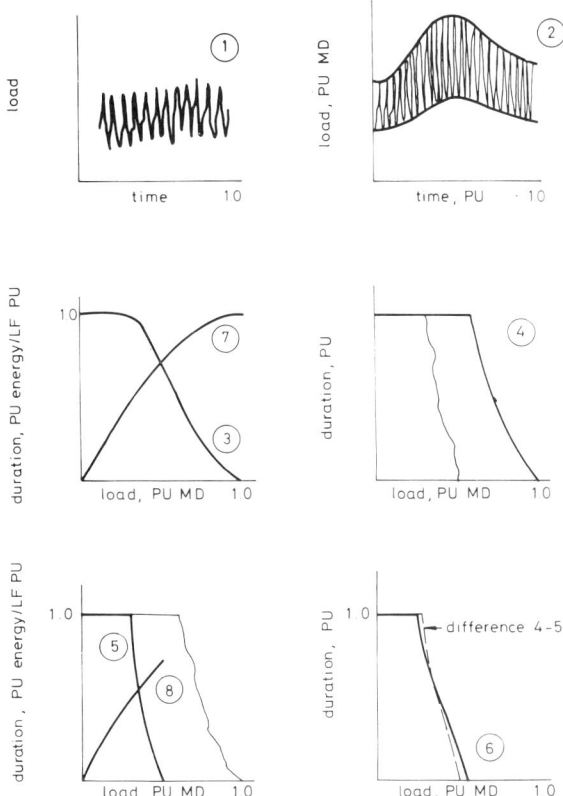

Fig. 5.2   System curves

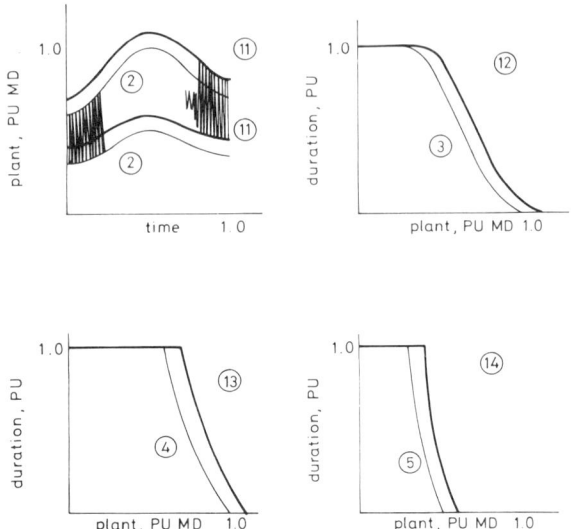

Fig. 5.3   Running plant curves

In order that the best use may be made of system load-duration curves and the various related, integrated and derived curves, it is important that the real significance of the co-ordinates of the curves should be recognised.   In particular, a clear distinction should be made between system requirement functions which are outside the system operator's control, and plant-use functions which show the operator's methods of meeting the imposed requirements with imperfect equipment.   These notes represent an attempt to define various curves and describe some of their uses.   A system of classification is proposed and then we give examples of the uses to which the curves may be put by a power-system planner.

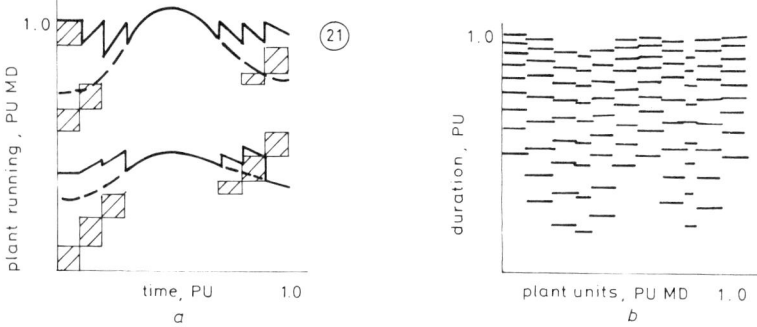

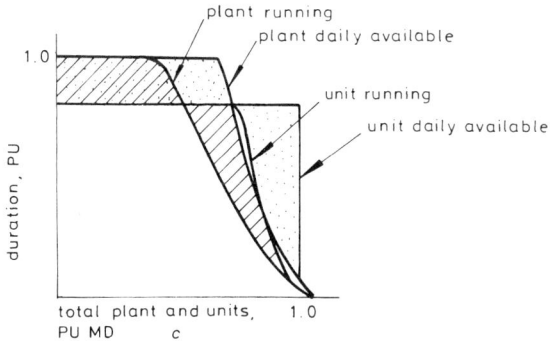

Fig. 5.4 Derivation of unit curves

    a: Running plant variation adjusted for maintenance outage
       = unit daily variation
    b: Maintenance outage as it would appear on a duration
       diagram

5.3.2 Classification

Curves which relate load and time in power systems fall naturally
into groups. The several categories of curves are indicated below
by the following classification symbols:

S   indicates system demands. This is most conveniently measured at
    power stations and tie-line terminals. In these cases it includes
    losses. It may or may not be adjusted for a suppressed component
    of electricity demand according to the purpose, but it is assumed
    to be outside the control of the system operator.

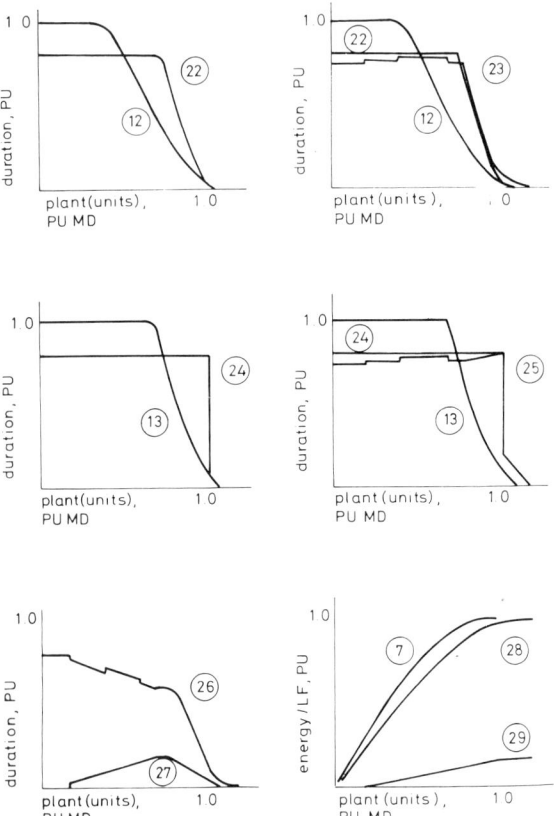

Fig. 5.5   Unit curves

P   indicates generating plant in terms of total aggregate rating.

U   indicates generating plant use in relation to the ratings of identifiable generator units, groups of units or plant categories. 'Use' in this sense includes running, loading, planned-outage for maintenance and forced outage.

T   represents real sequential time.

D   represents duration or the sum of elapsed time intervals. It can also be regarded as an expectation or, when dimensionless, as a probability.

Other curves which are sometimes useful relate electrical energy (kWh) production, or non-production, to system demand (kW), aggregate running plant and generating units. The following additional symbols are used to classify these curves:

E   represents cumulative energy (kWh) required by the system or supplied by the generating plant.

NE indicates unrealised energy (kWh). This is a concept used for the energy which could be, but is not, supplied by generating plant being only partly loaded, usually to provide spinning reserve.

In the illustrations, T is treated as an independent variable and plotted along the x axis against a vertical scale of power rating (kW). D is treated as a dependent variable and related to a horizontal scale of power or power rating (kW). This somewhat unusual way of presenting the parameters has been chosen to reduce the risk of confusion. Energy (kWh) is normally plotted on the same axis as T or D.

Adequate representation of a power system may require several curves in each category. The names are used to identify them and their general forms are indicated by Figs. 5.2 – 5.5. The diagrams are drawn on the assumption of a constant spinning reserve in all conditions, infinitely divisible generating capacity and, except in forced-outage conditions, strict merit-order running. By a constant spinning reserve we mean a spinning reserve related directly to the rating of the largest generating unit.

5.3.3 Construction of curves

Given a set of hourly system loading readings covering a whole year, or other periods of interest, curves (1) to (6) can be drawn directly by judicious selection and organisation of the data. It should be noted that the load duration curve (6) is not the difference between monotonic maximum and monotonic minimum except in the special case when both become monotonic for the same sequence of ordinates. It

can however be derived from either curve (4) or curve (5). Curve (7)
can be drawn by measurement of areas under curve (3) and curve (8)
by measurement of areas under curve (5).

5.3.4   Some useful curves

	Curve description	Class	Derived from
(1)	Daily load	ST	Records SR
(2)	Daily load variation	ST	Records SR
(3)	System load duration	SD	(1)
(4)	Daily or weekly peak load	SD	(2)
(5)	Daily or weekly minimum load	SD	(2)
(6)	Daily or weekly load range duration	SD	(2)
(7)	System load energy	SE	(3)
(8)	System block load energy	SE	(5)
(11)	Running-plant variation	PT	(2) + SR
(12)	Running-plant duration	PD	(3) + SR
(13)	Running-plant daily or weekly maximum	PD	(4) + SR
(14)	Running-plant daily or weekly minimum	PD	(5) + SR
(21)	Unit daily or weekly variation (running and planned outage)	UT	(11) + Maintenance programme
(22)	Unit running duration (without forced outage)	UD	(12) + Maintenance
(23)	Unit running duration	UD	(22) + F.O.R.
(24)	Unit required daily or weekly availability (without forced outage)	UD	(13) + Maintenance
(25)	Unit required daily or weekly availability	UD	(24) + F.O.R.
(26)	Unit equivalent F.L. duration	UD	(12) + (22) + op. policy
(27)	Unit equivalent N.L. duration	UD	(12) + (22) + op. policy
(28)	Unit energy	UE	(26)
(29)	Unit unrealised energy	UNE	(27)

Note: This list is not exhaustive and other curves might be required
        for special purposes.

Curves (11) to (14) require a knowledge of the amount of spinning reserve capacity to be added to system load to give total running plant. The upper and lower boundaries of curve (2) can be moved accordingly to give curve (11) and the derived curves (10) and (11), whether or not spinning reserve is the same at daily maximum and minimum loads and whether or not it shows a seasonal variation. If, and only if, it is constant throughout the period can it also be added to curve (2) to give curve (12). In any other conditions curve (12) can only be drawn by adding spinning reserve to the daily load curve (1) and reshuffling the ordinates to obtain a monotonic curve.

To proceed from total running generating plant requirements to utilisation of individual generating units it is necessary to allow for planned and forced outages. Since planned outage is by definition associated with a timetable it can be added directly to a sequential time diagram, but not to a duration diagram. Forced outage, until it occurs, must be regarded as a probability and treated stochastically. Therefore, it can be shown directly on a probability (or duration) diagram, but not on a sequential time diagram.

In Fig. 5.4a the diagram of running plant variation (11) has been modified to show planned outages of units and the resulting increase in daily use of lower merit plant-curve (21). A unit daily (or weekly) required availability curve (24), ignoring forced outage, can be derived. Similarly, a unit running duration curve (22) ignoring forced outage can be derived from (12) with a knowledge of planned outage.

Curves (22) and (24) can easily be modified to take account of the reduction in unit running times due to forced outage. However, the redistribution of load will depend on the type and state of readiness of reserve plant, on operating policy, and on chance, i.e. coincidence of forced outages. In the illustration the extra duty has been spread evenly between non-running plant and the daily required availability adjusted accordingly. This assumes the unlikely condition that spinning reserve is restored to the nominal value.

Given a knowledge of operating policy, the unit running curve can be subdivided into a unit equivalent full-load curve (26) and a unit equivalent no-load curve (27). The illustrations are based on a spinning reserve always 10% of maximum demand and spread over a plant rating of 40% of maximum demand; i.e. as much plant as necessary, starting with the lowest merit, is loaded to 75% of rating, any remaining higher merit running plant being on full load. It should be noted that these curves are equivalent to the true loading curve in respect of energy production but not necessarily energy cost, which will depend on the actual Willan's line being operated to on the thermo-dynamic chart.

Integration of the equivalent full-load and no-load curves (26) and (27) gives the unit energy curve (28) and the unit unrealised energy curve (29).

ACKNOWLEDGMENT

This Appendix, including the Figures, is the work of the late Mr. A.P. Coleman whilst with Preece, Cardew & Rider, Consulting Engineers, Brighton, UK, and is reproduced by permission of that firm.

REFERENCES

[1]   JONAS, P.J.: 'Fuels and generating plant mixes'. Paper given at International Symposium on Electricity Economics and Load Management, Imperial College, London, 20-24 Mar. 1980

[2]   BERRIE, T.W.: 'Applying economies of scale to generating plant'. Elect. Rev., 15 July 1977, pp 25-26

[3]   BERRIE, T.W.: 'Development of generating plant mix'. Elect. Rev., 22 Sept. 1967

[4]   'Central Electricity Generating Board'. The Monopolies and Mergers Commission, HC 315, HMSO, London, 20 May 1981, Chap. 7

[5]   House of Commons First Report from the Select Committee on Energy Session 1980-81, HC 114, 13 Feb. 1981; Monopolies Commission Investigation into the Central Electricity Generating Board, May 1981

[6]   TURVEY, R.(Ed.): 'Public enterprise'. (Penguin Economics Classics, 1968), Chap. 6

[7]   POSNER, M.: 'Fuel policy'. (North Holland Press, 1974)

[8]   BERRIE, T.W., and BETTS, P.: 'Assessment of costs of alternative plant proposals on the CEGB system'. UN International Atomic Energy Agency Conference, London, 1967

[9]   In the case of hydro-electric power stations, the amount of electrical energy output is limited. Their operation is usually reserved, therefore, for the time of system peak, when the generating stations which are most expensive to run would otherwise be in operation

[10]   This is usually laid down by Government for electricity generating projects. It applies to sums done in constant price levels (i.e. without inflation)

[11]   BERRIE, T.W.: 'Economics of system planning in bulk electricity supply'. Elect. Rev., 18 Sept. 1967. Strictly speaking, Fig. 1 shown therein shows a plant load factor curve in the upper diagram against a system load-duration curve in the lower

diagram. As shown in Appendix 5.3 we should convert our
load-duration curve into a plant-duration curve to make them
compatible

[12] TURVEY, R., and ANDERSON, D.: 'Electricity economics'.
Published for the World Bank (Johns Hopkins Press, Baltimore,
USA, 1977)

Chapter 6

Energy sector modelling

6.1 BACKGROUND TO MODELLING

6.1.1 Introduction

The low price of fuels up to 1973 encouraged the use of energy intensive processes in industry[1] together with an exorbitant use of fuels in the commercial and domestic sectors. In most countries the existing energy supply systems for coal, oil, gas and electricity have grown up independently and often competitively, and nowhere in the world can there be found a truly co-ordinated, integrated national energy supply sector.

We are becoming accustomed to frequent increases in energy prices resulting from rising fuel costs. In their turn, energy price increases influence the prices of all commodities, food and services according to individual energy intensities, and hence influence the standard of living and the Gross Domestic Product (GDP).

The elasticity of demand with respect to rising prices brings about an overall reduction of demand for energy and commodities, leading also to a reduction in the growth rate of GDP. Therefore, in present circumstances a simple extrapolation of past trends is inadequate for predicting the future demand for any form of energy, and some attempt must be made to model both the effect of general overall price increases and the reduction in demand for energy and goods associated with these increases. The kind of model which we use is that in which the characteristics of the major sectors of the economy, particularly energy, are expressed in mathematical form and operated on a digital computer.

In many parts of the world studies of future energy demand and supply are currently being made in a similar manner to that described in Chaper 1; these are being carried out in order that long-term plans can be formulated and new energy strategies discussed both nationally and internationally. A comparative survey and critique of the following nine world energy models most relevant to our approach to energy modelling has already been made by others[2]:

1   World Energy Conference (WEC)[3]
2   MIT: Workshop on Alternative Energy Strategies (WAES)[4] [5]
3   Nordhaus Model[6]
4   CIA Report[7]
5   OECD[8]
6   Ford Foundation Energy Project (F F EP)[9]
7   Nuclear Energy Policy Study Group (NEPSG)[10] [11]
8   Caltex Corporation
9   Exxon Corporation

Only a minority of these models (notably MIT, F F EP and NEPSG) make an attempt to fully integrate supply and demand through a simulated market mechanism.

In this Chapter we outline a methodology for projecting national energy demand and supply over a 30-year period. We describe a dynamic model designed as a 'test-bed' on which to investigate mainly the effects of various actions of government on national energy demand using different taxation/subsidy policies, because it is government action which dominates all energy scenes today. In such a model it is necessary to simulate all main aspects of consumer behaviour, because saving money and energy in one sector can merely lead to an equivalent expenditure in another. The model must trace energy flows from primary production, through industry and commerce to the consumer. Such a model has been used by Rodriguez[12] to produce various energy demand projects for the UK over a 30-year period.

6.1.2   Need for long-term demand projections

Planning, building and commissioning any major plant in the energy sector, a power station for example, often takes a whole decade. A typical timetable is given in Table 6.1.

Nowadays some would regard the time shown for each stage in Table 6.1 as optimistic, considering the recent record for power station construction in the UK. We must also remember that stage 1 can commence only after a system planning decision has been reached: in the case of Table 6.1 that a new power station should be constructed at a particular location. On today's evidence it takes about 6-10 years to bring a newly designed energy-conversion plant into operation. Therefore we must make long-term as well as short-term projections of energy demand. A typical horizon is some 30–40 years ahead because energy-conversion plants often have a serviceable lifetime of the order of 30 years.

Table 6.1 Problems with new plant

Stage	Problems	Time Years
1	Environmental objections Includes public inquiries, protest, demonstrations, etc. e.g. UK, France, W. Germany	1-2
2	Labour disputes on site Problems of organisation of large groups of construction workers: trade union disputes over pay and demarcation, management of sub-contractors. Budget control problems: inflation and high interest rates	5-7
3	Advanced technology New construction techniques for civil and mechanical engineering; advanced electrical machine design etc., e.g. AGR construction delays at Dungeness B	
4	Commissioning Teething problems of new technology. Manning of plant incorporating new process control techniques, automation and computers: unions in the UK will not normally commence negotiations over pay and conditions until plant is completed.	1-2

Total time: 7-11

Source: SHORT, M. J.: 'National energy supply systems – a model for demand projections'. Paper given at International Symposium on Electricity Economics and Load Management, Imperial College, London, 24–28 Mar. 1980. Reproduced by permission of Imperial College, Power Systems Group

6.1.3 National organisations making energy projections

In most countries today many national organisations need to make energy projections. For example, in the UK 12 organisations regularly issue their own energy demand forecasts, the most important being:

    Atomic Energy Authority
    British Gas Corporation
    British Petroleum
    British Steel Corporation
    Central Electricity Generating Board

Department of Energy
Electricity Council
Esso Petroleum (UK)
Imperial Chemical Industries
National Coal Board
Shell International
Shell (UK)

In North America, energy demand forecasts are regularly issued by government departments, the oil and major coal companies, some power utilities and academia. Throughout the world it is commonplace for large organisations to err on the 'safe' side and to over-estimate energy demand. Energy-sector planners have developed an exponential-growth type of mentality. Since they tend to take for granted that the trend in energy demand is increasing exponentially, the only factor over which disagreement then arises is the percentage annual

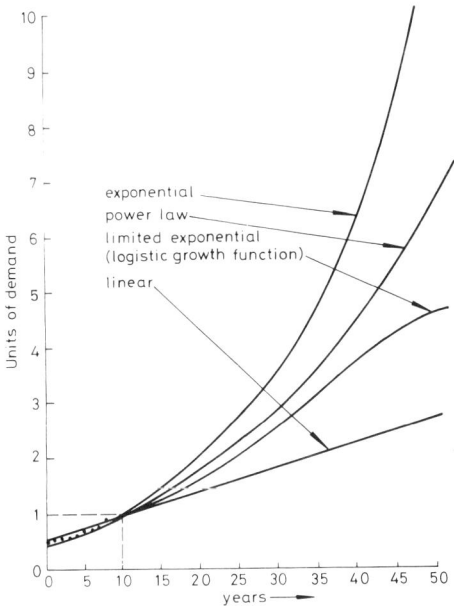

Fig. 6.1   Extrapolation of past trends

SHORT, M.J.: 'National energy supply systems – a model for demand projections'.   Paper given at International Symposium on Electricity Economics and Load Management, Imperial College, London, 24-28 March 1980. Reproduced by permission of Imperial College, Power Systems Group

growth rate. Comparatively recent 'official' UK projections of annual growth rates have ranged from under 3% (Department of Energy) to over 5% (UKAEA) - a spread of about 2 to 1.

The effect of a simple extrapolation of the trend in the past 10 years is indicated in Fig. 6.1. The uncertainty of the projection 30 to 40 years ahead can easily span an order of magnitude: bounded at the upper extreme by the exponential projection and the lower extreme by a linear projection. A logistic growth function would seem to be more reasonable in the light of past experience, except in the smaller developing countries where exponential growth rates might still give the right projections.

In 1976 the Open University Energy Research Group[13] commented that the grossly exaggerated prediction for UK electricity demand over the preceding decade was largely because the forecasters had failed to allow for the rapid penetration of North Sea Gas into electricity markets together with the increasing saturation of many domestic markets for electricity. Prior to North Sea Gas, the British gas industry had been moribund.

Evidence of saturation of a market is indicated by trends such as those in Fig. 6.2a (cars in the UK) and Fig. 6.2b (UK output of electricity). In both cases the rates of change per annum are more significant than the absolute levels, since a downward trend in a rate of change indicates the onset of saturation. Once a quasi-linear region in a logistic growth function (see Fig. 6.2c) has been detected, then both the time taken to reach saturation and the saturation level itself become parameters of great significance to the energy planner. Only with a full dynamic simulation of the national energy-supply/demand system through the market mechanisms can the likely logistic growth functions be determined for a given range of scenarios.

6.2 RECOMMENDED ENERGY MODEL STRUCTURE[14]

6.2.1 Methodology

A model for national energy demand projections[15] taking into account interaction of supply and demand through this market mechanism should include the following inputs:

(i)   gross domestic product (GDP)
(ii)   energy prices
(iii)   population
(iv)   number of households
(v)   structure of households
(vi)   fuel tax
(vii)   government subsidies for insulation materials
(viii)   government subsidies for public transport
(ix)   conservation time constant
(x)   exports

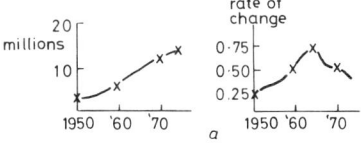

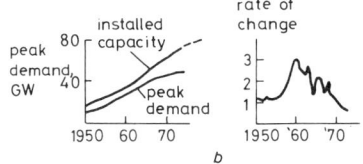

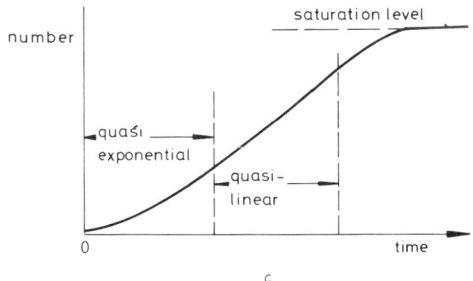

Fig. 6.2   Saturation of demand

     a   Cars in the UK
     b   Output of electricity
     c   Logistic growth function

     SHORT, M.J.: 'National energy supply systems – a model for demand projections'. Paper given at International Symposium on Electricity Economics and Load Management, Imperial College London, 24–28 March 1980. Reproduced by permission of Imperial College, Power Systems Group

(xi)   imports

The output of the model includes both money and physical demand for all goods disaggregated in the model, including the various energy sectors, and also the extent that various energy-conservation measures have been taken up. We can then estimate how consumption patterns and conservation behaviour have been affected by energy price changes and government action.

Our model runs start in a chosen base year; they proceed, recalculating all the variables for each time-step given until the end of the computer run. Calculations performed in each time-step include:

(i)   adjustment of any exogenous variable as necessary from input data

(ii)   recalculation of prices, making use of an input/output table (see later)

(iii)   optimal points for capital/energy substitution

(iv)   new optimum fuel mix (domestic and industrial)

(v)   final consumption calculated for each commodity group, using economist demand curves and price elasticities

(vi)   final demands adjusted for imports and exports

(vii)   input/output matrix inverted to give gross demands for all fuels and commodities (see later).

6.2.2   Model details

At Imperial College, London, a dynamic energy model for the UK has been made on the above lines,[12] by dividing the national economy into 24 sectors, five of them representing energy (coal, oil, petrol, gas and electricity). The 24 sectors were considered to be the minimum number to represent the UK and still retain a 'feel' for the economy. The sectors are arranged in five groups as shown in Table 6.2.

Several 'modules' are included in the computer program to represent the factors mentioned above of capital/energy substitution, fuel mix and substitutability according to relative fuel prices, technological progress, transport modes etc. The modular structure of the program is shown in Fig. 6.3.

The step-by-step dynamic simulation proceeds in the model by forming and integrating a series of 'state' equations, i.e. a first-order time derivative of a 'state' variable. One example is the equation for commodity prices:

$$\frac{d}{dt} SPRICE(J) = \frac{1}{TPRICE(J)} \cdot [(PRICE(J) - SPRICE(J))]$$

where $SPRICE(J)$ = smoothed price of the J th commodity

$TPRICE(J)$ = price-smoothing time constant, i.e. time for a price change to 'work through' the economy, say 6 months

$PRICE(J)$ = current price of commodity J .

Table 6.2   24 Sectors in five groups

Group No.	Category	Sector No.	Title
1	Energy:	1	Coal
		2	Oil
		3	Petrol
		4	Gas
		5	Electricity
2	Materials:	6	Iron and Steel
		7	Other metals
		8	Chemicals
		9	Construction materials
		10	Construction
		11	Capital goods
3	Consumer goods:	12	Food
		13	Drink and tobacco
		14	Textiles and clothing
		15	Non-electrical durables
		16	Electrical appliances
		17	Paper and non-durables
		18	Motor vehicles
4	Transport:	19	Road haulage
		20	Buses and taxis
		21	Trains
		22	Air/sea
5	Services and Government:	23	Services
		24	Public authorities

Source: SHORT, M. J. 'National energy supply systems – a model for demand projections'.   Paper given at International Symposium on Electricity and Load Management, Imperial College, London, 24-28 March 1980.   Reproduced by permission of Imperial College, Power Systems Group

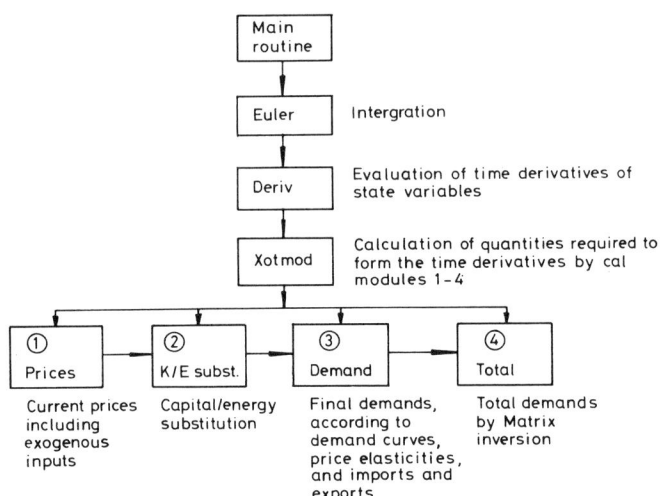

Fig. 6.3   Modular structure of dynamic simulation program

SHORT, M. J.: 'National energy supply systems – a model
for demand projections'. Paper given at International
Symposium on Electricity Economics and Load Management,
Imperial College, London, 24–28 March 1980. Reproduced by
permission of Imperial College, Power Systems Group

The current prices are calculated using a national economic
input/output table.[16,17] The principle of an economic input/ output
table is illustrated in the simple example given in Table 6.3. All
entries in the table are in money values. The input/output (I/O)
table enables a matrix of I/O coefficients to be calculated and the
prices module in Fig. 6.3 operates as shown below:

Exogenous Fuel Prices

↓

Smoothed prices of all   →   Matrix of I/O   ←   Residual price terms
        commodities   Coefficients   (including trends)

↓

Current Commodity Prices

Capital/energy substitution (module 2 in Fig. 6.3) operates by
calculating instantaneous optimum points, given the current prices
of energy. Capital movement towards these optimum points is then
made subject to time delays.

Industrial and domestic fuel mix, and substitutability of fuels,

Table 6.3   Example of a simple input/output table

	Sector 1	2	3	4	Intermediate output	Output to final consumer	Gross output
Sector 1	0	0	10	20	30	60	90
Sector 2	2	0	20	2	24	0	24
Sector 3	10	8	0	5	23	150	173
Sector 4	10	5	20	0	35	5	40
Intermediate Inputs	22	13	50	27		↓	
Value added	68	11	123	13	→	215 †	

Columns 1-4 show <u>inputs</u> to sectors

First four rows show <u>outputs</u> of sectors

†Total Value Added = total final consumption expenditure = gross domestic product (GDP)

SHORT, M. J.: 'National energy supply systems – a model for demand projections'. Paper given at International Symposium on Electricity Economics and Load Management, Imperial College, London, 24-28 March 1980. Reproduced by permission of Imperial College, Power Systems Group

take account of relative fuel prices and incomes, incorporating the appropriate elasticities.[18],[19]
Final demands for commodities are calculated according to demand curves,[20],[21] allowing for household income, price elasticities and imports and exports. The total demands for all commodities are calculated by matrix inversion, using Leontief's method.[22]
In the Imperial College model projections of the UK demand for electricity have been examined specially. All the mechanisms described above were included and a range of possible output trajectories was obtained using a range of plausible input scenarios; see Reference [12].

6.3   Energy Model Applications

Some potential applications for dynamic energy demand models like the above are:

(a) <u>Projection of national demand for energy</u>, i.e. at least 10 to 20 years and probably further ahead, with adequate attention to uncertainties, so that a <u>range of possible output trajectories</u> is produced. Effects that can be studied include those of:

(i)   rising fuel prices
(ii)   government tax and subsidy policies
(iii)   feedback of rising prices on demand and on rate of growth of GDP
(iv)   technological progress
(v)   balance of imports/exports.

Outputs will include the gross demands for coal, oil, gas and electricity.

(b) <u>Energy balance in a nuclear-based economy</u>; i.e. an additional sector for nuclear power can be included in the model having the major inputs of energy, materials, construction and labour and producing useful electrical energy after, say, a 10 year delay. The energy balance and costs in the long run can be assessed.

(c) <u>National transport policy</u>; i.e. the effects which can be studied include those of:

(i)   taxation/subsidy policies on the private-car/public transport balance
(ii)   widespread use of electric vehicles and electric traction in place of petrol/diesel vehicles

(d) <u>International energy projections</u>, i.e. for EEC, OECD, developing countries. Each country, or group, can be modelled and interconnected via import/export matrices.

(e) <u>Multi-national corporations and private companies</u>, i.e. for future planning and formulation of long-term strategies with respect to energy.

RE F ERENCES

[1]   SHORT, M. J.: 'National energy supply systems – a model for
       demand projection'. Paper given to the International Symposium
       on Electricity Economics and Load Management, Imperial College,
       London, 24-28 March 1980. This chapter is based upon this paper

[2]   See especially ULPH, A.M.: 'World energy models – a survey and
       critique', Energy Economics, J an. 1980, 2, (1), pp. 46-59

[3]   World Energy Conference: 'Study Group Report on World Energy
       Demand and Study Group Report on Oil and Gas Resources' (IPC
       Science & Technology Press, Guildford, UK, 1978)

[4]   Workshop on Alternative Energy Strategies: 'Energy: Global
       Prospects 1985-2000' (McGraw-Hill, New York, 1977)

[5]   Workshop on Alternative Energy Strategies: 'Energy supply-
       demand integrations to the year 2000: Global and national
       studies' (MIT Press, Cambridge, MA, 1977)

[6]   NORDHAUS, W.D.: 'The allocations of energy resources'.
       Brookings Papers on Economic Activity, Vol. 3, 1973, pp. 529-576

[7]   CIA: 'The international energy situation outlook to 1985'.
       CIA-ER77-10240 V, Washington DC, 1977

[8]   OECD: 'Energy outlook to 1985'. OECD, Paris, 1977

[9]   Ford Foundation Energy Project: 'A time to choose' (Ballinger,
       Cambridge, MA, 1974)

[10]   Nuclear Energy Policy Study Group: 'Nuclear power issues and
       choices' (Ballinger, Cambridge, MA, 1977)

[11]   MANNE, A.: 'ETA – a model for energy technology assessment'.
       Bell J. Economics, 1976, 7, (2), pp. 379-406

[12]   RODRIGUEZ, A.: 'Economic model for energy planning'. M.Sc.
       Report, Imperial College, London, 1978

[13]   Energy Research Group, Open University: 'A critique of the
       electricity industry'. ERG 013, March 1976

[14]   DANSKIN, H.: 'SARU'S energy-demand model', Futures, Dec. 1979,
       pp. 491-509

[15]   Departments of Environment and Transport: 'UK Energy demand
       projections'. Aug.1979

[16] CSO: 'Input-output tables for the UK, 1968'

[17] '1972 input-output tables of the UK'. PA1004 Business Monitor
      CSO

[18] Department of Energy: 'Report of working group on energy
      elasticities'. Energy Paper 17, 1977

[19] Department of Energy: 'Energy balances - some problems and
      recent developments'. Energy Paper 19, 1977

[20] Department of Employment: ' Family expenditure survey Report for
      1972' (HMSO, 1973)

[21] Departments of Environment and Transport: 'Sarum 76 - Global
      Modelling Project'. Research Report 19, 1977

[22] LEONTIEF , W.W.: 'The world economy of the year 2000', Scientific
      American, Sept. 1980, 243, (3)

Chapter 7

Power sector modelling

7.1 BASIC APPROACHES

7.1.1 Introduction

Much effort has gone into mathematical models for economic planning and operation of power systems. Details of some models are given below and in Appendix 7.4. We only deal here with simple illustrations from a vast volume of published material[1], and mainly with planning. Power system planning is mainly concerned with answering:

When should plant be installed?

Where should plant be installed?

What type of plant should be installed?

What size of plant should be installed?

We cannot answer these questions properly unless we know the answer to the fundamental question: what is the prime objective(s) of power system planning? We answered this by defining the investment rule in Chapter 1. This rule is at the heart of all system planning models, with the reliability rule also playing an important part.

However, no matter how true this is of our model and no matter what its sophistication, it will never capture everything in the decision-making process of system planning, even though more explicit indicative planning can be done with power-sector models than with energy-sector models.

Power-system models are basically concerned with optimisation, usually in finding the least-cost development plan for the power utility. If we accept the planning objective in this form we must state the constraints which the utility must satisfy when trying to achieve its planning objective, some imposed from within the utility itself, others imposed from outside, mainly by the economy or government:

(i)   reliability constraints
(ii)   economic constraints
(iii)   technical constraints
(iv)   environmental and social constraints
(v)   legal constraints
(vi)   political constraints.

Each system planning exercise has its own mixture of these constraints plus some constraints unique to the occasion. The relative importance of these constraints will affect the form which planning models will take, although all system planning models have a basic approach which we will now describe.

7.1.2 Basic approach to power-sector modelling: decomposition

The basic approach to a complex planning situation in mathematical terms is a process normally called 'decomposition', now described in a simplified way. One form of simple decomposition is the disaggregation of the national economy into the sectors of Chapter 6. In other forms of decomposition we decompose problems according to time, i.e. long, intermediate and short-term problems, or according to function, e.g. technical, operational, economic and financial.

A simple framework describing functional decomposition is shown in Fig. 7.1. We show no arrows, to emphasise that the process of planning is really circular and not linear, i.e. includes iterations. However, it is convenient to break into the planning circle at the load forecast stage and this is the basis for drawing Fig. 7.1.

Starting at the top of Fig. 7.1 with forecasts of peak power (kW) and electrical energy (kWh) for future time periods, we have a first shot at determining the best system expansion plan, involving new generating and network plant. At this stage the first answers to the above questions of when, where, type and size of plant can be spelled out.

A variety of models exist to determine optimum system expansion plans, basically using analysis rather than synthesis, i.e. analysing a plan put together by a separate process.

After the expansion-plan models in Fig. 7.1 come the network-analysis models for checking load-flows, short-circuits, stability, reliability etc. These form part of the technical analysis of a proposed development programme and the reliability analysis of the system components.

Following the technical analysis, Fig. 7.1 goes in two different directions. In one, by simulating future system operation, we assess the expansion plan in terms of system electrical energy production

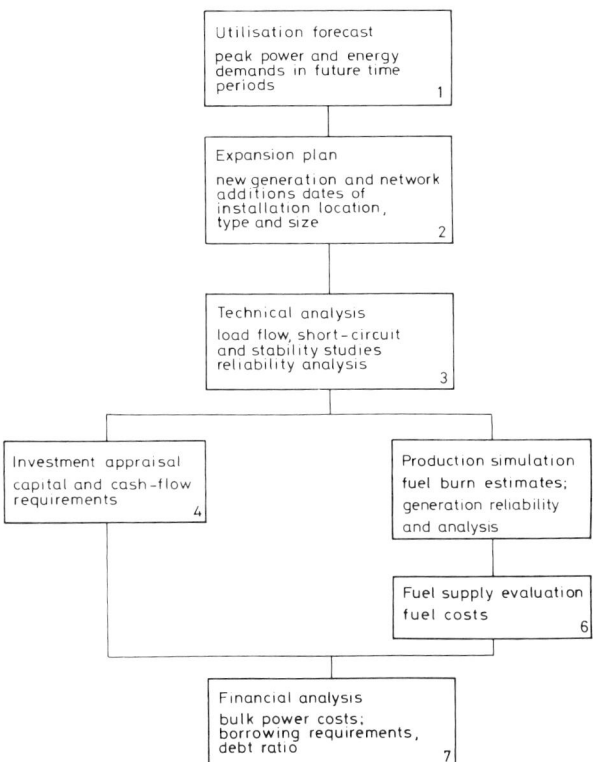

Fig. 7.1   Stages in power system planning

LAUGHTON, M.A.: 'Electrical power systems - Models for
expansion planning'. Paper for International Symposium
on Electricity Economics and Load Management, Imperial
College, London, 24-28 March 1980. Reproduced by permission of Imperial College, Power Systems Group

and fuel burn. Again, many models are available to do this. Indeed,
models are needed for day-to-day system operation, mostly developed
by the utilities themselves, and fully acceptable for the overall
planning and operation of all power utilities.[2] Using such models
we look at the system capability to supply electrical energy, when,
where and to what extent it is required, and also at the optimum and
actual reliability of supplying it (Chapter 3). Having simulated
the system we determine the fuel requirements and the type and sources
of fuels; and fuel costs.

Feedbacks exist, although for simplicity are not shown in Fig. 7.1.

For example, if we put together an expansion plan and it fails in technical analysis there will be a feedback to change the plan. Equally, if the expansion plan fails to satisfy the required production function there will be a feedback signal. The same is true for limitations in the fuel supply. In reality we have in a working model, of which Fig. 7.1 is a very simplified version, a network feeding back information and feeding forward modifications.

On the second side of the split in Fig. 7.1 we have investment appraisal in which the capital costs in every year of the development plan are joined with the attributable yearly running costs of the whole system, this being done over the economic life of the plan, using some type of discounted-cash-flow techniques to determine a total present value (Chapter 8). We must also look at cash-flow requirements with respect to the utility's borrowing needs and its debt-service position.

At the bottom of Fig. 7.1 we are in the utility's financial controller's office with access to all information required for calculating the long-run supply costs from which we can set tariffs. From these tariffs we can test whether our electricity load forecast should be modified because electricity prices originally assumed were different from the long-run supply costs just determined, thus giving rise to another iteration through Fig. 7.1

7.1.3 Power-system modelling

Some utilities use very sophisticated models.[3],[4] As this is not a book on modelling, we describe only some common modelling methods; first returning to Fig. 7.1 to discuss a general but important point for all power-system models.

When using the models that fit into boxes 1-3 in Fig. 7.1, we would like to know the problems likely to arise in boxes 4-7 'downstream'. Fortunately experience has shown that it is not necessary to fully simulate the 'downstream' problems at this stage; in boxes 1-3 we deal at first only with a best-guess plan, evaluating the 'final' long-term expansion plan later and possibly after several iterations through Fig. 7.1 via pricing. The iterations gradually bring in the 'downstream' effects until an overall optimum for all components of Fig. 7.1 has been reached.

The method chiefly recommended when optimising a non-technical box in Fig. 7.1 is linear programs (LP). The applicability of LP to system planning is examined extensively in the literature[5]; here we look only at those features of LP most directly related to system planning.

A sound reason for using LP in system planning is that any computer manufacturer or 'software' specialist will offer a package of programs which take in LP problems as their input, solving the problems and producing answers in a standard output format. Into most LP packages has gone a programming manpower effort of at least 15 to 20 man-years. Thus we have immediately available off-the-shelf sophisticated programs for our own particular optimisation analyses.

All the planner has to do is to put his particular optimisation problem
into a form which his LP package can handle. Although we may still
have to write a program to control the sequence of calculations within
the LP package or between LP packages, the main task is to get our
problem into a form in which it can be readily solved by LP methods.
   Fortunately any simple problem can be stated in an LP form. First
we specify an objective function which includes all the necessary
attributes of the problem which the model needs to know. For LP this
objective function must be expressed in a linear function;
constraints on this objective function must also be stated in a linear
manner. We solve our problems by successive applications of LP.
To illustrate this, we now go through a simple exercise in system
planning by successive application.

7.1.4 Simplified example

   Let us plan to build only two power stations A and B, knowing that
B is more efficient than A; every input to B produces 20 units of
output, the same input to A produces 10 units of output; i.e. 20 units
of B are produced for 10 units of A. There are other constraints;
e.g. the output of A plus B must at any instant equal total demand
on the system neglecting plant margins and there will be some
constraints in system operation. Now the boundaries forced by
constraints on the objective function can move around only within
the strict confines of the boxes of Fig. 7.1. As one constraint is
activated after another, the objective function is satisfied by
different values of output capacities of generating stations A and
B until we reach the overall optimum, i.e. at those values for which
it is difficult to move around within the constraints and within the
boxes of Fig. 7.1 and still find a better way of satisfying the
objective function. Let us say that the overall optimum values for
the output capacities from power stations A and B are found by
successive approximations to be 50 MW and 150 MW, respectively, for
meeting a peak load of 200 MW. Fortunately in practice only a few
constraints are really important.
   After finding the optimum solution we must carry out sensitivity
analysis. For example, we must ask what happens to the optimal
solution above if: (i) the amount of capital available changes; (ii)
the demand for electricity changes; and/or (iii) other important
constraints change? Also what happens if the capital costs of our
generating units turn out to be different from what we have assumed?
Does it make a difference to our optimal solution between A and B
above?
   Let us look again at the two power stations A and B making up a
power system. If we vary some of the constraints, then the overall
optimal solution will be insensitive to changes in some constraints,
possibly fuel availability; making more coal available in Britain
does not mean that more coal-fired generating stations should
necessarily be built. However, the overall optimum might be very

sensitive to changes in value of some constraints, in which case the decision pattern will move away from the 50 MW of A and the 150 MW of B to some other balance. A change in capital cost of B might, for example, mean 100 MW of each station.

Now, if generating station A becomes more efficient, then the original 50/150 ratio of output capacity A to B will change. Normally we are not looking at a continuous phenomenon because, as more efficient plants are being added to the system, quite suddenly the original plant mix is 'wrong', and our model must be able to tell us when this change has occurred.

7.2 A WORKED EXAMPLE

7.2.1 Introduction

Using the reserve plant margin approach to reliability the available capacity of the generating plants times the number of generating plants must be greater than or equal to peak demand plus the reserve margin, taken here as 20%.

The peak demand constraints are thus:

$$\sum_{v=0}^{t} \sum_{j=1}^{J} a_{jv} x_{jv} \geqslant d_t(1 + mx), \text{ for } t = 1 \ldots\ldots T \tag{7.1}$$

where T = number of intervals in the period of the study

a_{jv} = available capacity of generating plant type j installed in time interval v, i.e. installed capacity times plant availability.

x_{jv} = number of plants of type j installed in interval v

d_t = peak demand in interval t

v = date of installation of a plant, i.e. v = 0 identifies existing plants

m = reserve margin as a percentage of d_t

J = number of different types of plant

The values of x must be integer; 1,2,3....n, where n is an integer greater than the number of plants likely to be needed at a particular time.

If we look at Tables 7.1 - 7.3 we will see the type of information which we usually need before we can start any planning study, e.g. the number of existing and possible future generating units of any particular type and capacity, forced outage rates, operating costs, maximum rate of introduction of new plant, investment costs of new plant, duration of load at any particular level of MW, i.e. the load-duration curve. Some information is historical, e.g. forced outage rates of existing plant. This includes information of the following type: what percentage time a particular plant was out of service; for a new plant, what percentage time we expect it to be out of service. The load-duration curve's shape is usually based on past experience, merely substituting new relative values for the future.

Table 7.1   Existing plants

Capacity, MW	No. of units	Forced outage* rate, %	Operating cost, £/MWh
50	4	5.0	17.0
100	5	5.0	9.2
100	6	6.0	9.0
200	5	6.0	8.5
400	2	8.6	7.7
600	1	12.0	3.2

* includes adjustment for maintenance

Table 7.2   Future plants

Capacity MW	Max. rates of* introduction	Forced outage† rate, %	Operating cost, £/MWh	Investment cost, £/10^6
200	8	9.8	16.0	35.08
400	4	9.1	8.0	120.00
600	4	11.8	7.0	152.00
800	4	12.4	7.0	185.08
1000	3	15.0	3.0	474.70
1200	3	20.0	3.0	531.90

* Used only for the integer solution
† Includes adjustment for maintenance

Source: LAUGHTON, M.A.: 'Electrical power systems – models for expansion planning'.   Paper for International Symposium on Electricity Economics and Load Management, Imperial College, London, 24–28 March 1980. Reproduced by permission of Imperial College, Power Systems Group

Table 7.3   Deterministic load-duration curve

Load level MW	Dura-tion h	Load level MW	Dura-tion h	Load level MW	Dura-tion h	Load level MW	Dura-tion h	Load level MW	Dura-tion h
3000	1	2700	12	2400	24	2000	360	1400	192
2975	1	2675	12	2375	48	1950	456	1350	168
2950	1	2650	24	2350	72	1900	696	1300	144
2925	2	2625	12	2325	48	1850	768	1250	108
2900	2	2600	24	2300	48	1800	912	1200	89
2875	4	2575	24	2275	48	1750	1032	1150	101
2850	3	2550	24	2250	72	1700	468	1100	144
2825	4	2525	36	2225	96	1650	372	1050	108
2800	4	2500	36	2200	96	1600	276	1000	48
2775	8	2475	24	2150	168	1550	264		
2750	8	2450	48	2100	192	1500	288		
2725	12	2425	48	2050	264	1450	216		

Peak : 3000 MW
Reserve as a percentage of peak : 20%
Annual rate of increase of load : 8%
Annual discounting rate : 10%
Period of study : 10 intervals of length 2 years
Post-horizon period (in which load
is considered to be constant) : 20 years

Source: LAUGHTON, M.A.: 'Electrical power systems − models for expansion planning'.   Paper for International Symposium on Electricity Economics and Load Management, Imperial College, London, 24−28 March 1980. Reproduced by permission of Imperial College, Power Systems Group

7.2.2   Importance of defining reliability standards

When we talk about a generating plant of size say 200 MW or 300 MW of available output, these figures represent the 'rated output' capacity, and not the capacity always likely to be available, because the plant sometimes will be out for maintenance and there may be unplanned outages. The degree of planned and unplanned outages give the 'forced outage rates'. Thus there is an 'effective' load-carrying capability of a generating unit, defined as the maximum increase in load sustainable by the system after adding the unit in question. From the figures in Table 7.1 and 7.2 on forced outage rates we can calculate the loss of load probability (LOLP) (Chapter 3). We use this as a measure of the overall reliability of the system. Because with an unreliable system we cannot count on as much capacity as on a more reliable one, more total capacity is needed to meet the same load forecast; e.g. if we add a 1000 MW unit to the power system at an assumed forced outage rate (Table 7.2) we can recalculate the reliability of the whole system when the 1000 MW unit has been added at (say) a loss of load probability of 2.4 hours per year, i.e. one tenth of a day per year, or one day every 10 years. If we now look at the difference with and without the 1000 MW unit, we will see below that we have actually added much less than 1000 MW, as system operators well know!

7.2.3 Detailed example

Using the model of Fig. 7.1 with a 20% reserve margin of generating plant starting from an arbitrary basis in box 2, we emerge via LP (and possibly integer programming)[6] with a development programme for installing generating plant over the 'planning period' chosen. Let us assume we start by adding 1000 MW of generating plant capacity. At the LOLP level of 2.4 hours per year we have added about 369 MW capability (Table 7.4). Since the 1000 MW set is assumed to have a forced outage rate of 15% (Table 7.2), we can thus count on 850 MW, i.e.
$1000 \times (1 - 0.15)$ of available capacity (Table 7.4). Carrying on with LOLP analysis we can then test the reliability level of the development programme started with in box 2 in Fig. 7.1. The results are given in Table 7.5.

The reliability of the simple system described in Table 7.4 varies considerably as shown in Table 7.5; only later is it anywhere near the 2.4 hours per year, our planning standard. Up to interval 7 the system is not reliable enough, and after interval 7 it is too reliable. From this test it appears that the planning plant reserve margin approach, i.e. the 20%, is not an accurate way to achieve a given reliability level.

If we modify the system demand function and build other things into the model such as a capacity constraint we can further develop the model used above.

The <u>capacity constraints</u> we will build in will look something like eqn. 7.2:

$$U_{jvtp} \leqslant a_{jv} x_{jv}$$

$$U_{jvtp} \geqslant 0, \forall_{jvtp}$$

$$0 \leqslant x_{jv} \leqslant n_{jv}, x_{jv} \text{ integer } \forall_{jv} \tag{7.2}$$

for $j = 1 \ldots J$, $v = 0 \ldots t$, $t = 1 \ldots T$, $p = 1 \ldots P$, where the symbols are the same as for eqn. 7.1 plus:

U_{jvtp} = level of production in sub-period p of interval t of plant
type j installed in interval v

n_{jv} = maximum number of plants of type j that can be installed in
interval v.

Table 7.4   Reserve margin approach in generation planning:
Capability study of an investment plan

Interval	Units added	Average capability, MW	Available† capacity, MW	ICCB*, £x10⁶/MW	ICCP**, £x10⁶/MW
1	1000	369	850	1.286	0.558
2	1200	403	960	1.300	0.554
3	3 × 200	208	180	0.169	0.195
4	1200	543	960	0.980	0.554
5	6 × 200	215	180	0.163	0.195
6	7 × 200	204	180	0.172	0.195
7	3 × 200	201	180	0.175	0.195
	1200	696	960	0.764	0.554
8	2 × 200	203	180	0.173	0.195
	2 × 800	693	701	0.267	0.264
9	200	196	180	0.179	0.195
	2 × 1200	790	960	0.673	0.554
10	2 × 800	718	701	0.258	0.264
	1200	838	960	0.635	0.554

†Available capacity = $(1 - FOR_j) \times CAP_j$

*ICCB:investment cost per unit of actual capability (not discounted)
**ICCP:investment cost per unit of available capacity (not discounted)

FOR = Forced Outage Rate
CAP = Rated Capacity
Investment cost   1817.87 (£×10⁶)
Production cost   1420.35 (£×10⁶)
Total cost   3238.22 (£×10⁶)
Lower bound   3178.31 (£×10⁶)
Precision   1.88%

Source: LAUGHTON, M.A.:'Electrical power systems – models for
expansion planning'.   Paper given at International Symposium on
Electricity Economics and Load Management, Imperial College, London,
24–28 March 1980. Reproduced by permission of Imperial College, Power
Systems Group

Table 7.5 Reserve margin approach in generation planning:
Reliability characteristics of the investment
plan of Table 7.4

Inter- val	Load, MW	Available capacity, MW	Actual capability, MW	LOLP (hours/year)
1	3499	4278	3198	8.44
2	4081	5238	3601	11.10
3	4761	5778	4225	11.59
4	5553	6738	4768	14.87
5	6477	7818	6058	6.55
6	7555	9078	7486	2.84
7	8812	10578	8785	2.55
8	10278	12340	10577	1.38
9	11988	14440	12353	1.26
10	13983	16802	14627	0.95

Source: LAUGHTON, M.A.: 'Electrical power systems - Models for
expansion planning'. Paper given at International Symposium on
Electricity Economics and Load Management, Imperial College, London,
24-28 March 1980. Reproduced by permission of Imperial College, Power
Systems Group

In this model plants of the same type have the same capacity, forced
outage rate, running costs and investment costs. With the capacity
constraints expressed as eqn. 7.2 we can examine a different pattern
of plant development over time. For example, instead of installing
one 1000 MW unit in interval 1 in Table 7.4 it might be better to
install 4 × 200 MW units (Table 7.6 refers). Thus the whole decision
pattern is shifted; it is sensitive to the assumed standard of
reliability with respect to the actual simulation of generating-plant
operation. Tables 7.6 - 7.8 refer to this stage of the model.

The original objective with respect to the reliability standard
was to keep the loss of load probability at about 2.4 hours per year.
Table 7.7 shows that, with this later analysis, using 4 × 200 MW
instead of 1 × 1000 MW, this has been achieved reasonably well and
certainly better than in the Table 7.5. The predicted capability
of generating units used in the model is shown in Table 7.8.

Table 7.6   Capability prediction model in generation planning:
            Investment plan

Inter-val	Units added	Average capability, MW	Predicted capability, MW	Actual ICC*, £x10⁶/MW	Predicted ICC*, £x10⁶/MW
1	4 × 200	194	200	0.181	0.175
2	1000	404	500	1.175	0.949
3	200	192	200	0.183	0.175
	1000	518	590	0.916	0.805
4	1000	623	650	0.762	0.730
5	2 × 200	211	200	0.166	0.175
	1000	691	690	0.687	0.688
6	200	208	200	0.169	0.175
	1000	727	720	0.653	0.659
7	3 × 200	208	200	0.169	0.175
	1000	748	750	0.635	0.633
8	7 × 200	203	200	0.173	0.175
9	5 × 200	200	200	0.175	0.175
	800	683	680	0.271	0.272
10	3 × 200	199	200	0.176	0.175
	2 × 800	693	690	0.276	0.268

*ICC - investment cost per unit of capability, either actual or predicted capability (cost not discounted)

Investment cost	1663.56 (£×10⁶)
Production cost	1570.86 (£×10⁶)
Total cost	3234.42 (£×10⁶)
Lower bound	3139.90 (£×10⁶)
Precision	3.01%

Source: LAUGHTON, M.A.: 'Electrical power systems - Models for expansion planning'.   Paper given at International Symposium on Electricity Economics and Load Management, Imperial College, London, 24-28 March 1980. Reproduced by permission of Imperial College, Power Systems Group

Table 7.7 Capability prediction model in generation planning:
Reliability characteristics of the solution in
Table 7.6

Inter-val	Load, MW	Predicted Capability, MW	Actual capability, MW	Difference MW	LOLP hrs/year
1	3499	3629	3605	24	1.38
2	4081	4129	4009	120	3.43
3	4761	4919	4719	200	2.78
4	5553	5569	5342	227	4.14
5	6477	6659	6455	204	2.57
6	7555	7579	7390	189	3.40
7	8812	8929	8762	167	2.66
8	10278	10329	10183	146	2.88
9	11988	12009	11866	143	2.79
10	13983	13989	13849	140	3.02

Table 7.8 Predicted capability of generating units in MW;
intervals since start up

Capacity, MW

Interval, years	200	400	600	800	1000	1200
1	200	360	410	430	390	320
2	200	370	450	500	500	460
3	200	380	480	560	590	560
4	200	380	500	600	650	630
5	200	390	520	630	690	680
6	200	390	530	650	730	720
7	200	390	540	660	750	750
8	200	390	540	680	770	770
9	200	390	550	680	780	790
10	200	400	550	690	790	800

These values are rounded up to their nearest multiple of 10
The total capability of the initial system is equal to 2829 MW

Source: LAUGHTON, M.W.: 'Electrical power systems - Models for
expansion planning'. Paper given at International Symposium on
Electricity Economics and Load Management, Imperial College,
London, 24-28 March 1980. Reproduced by permission of Imperial
College, Power Systems Group

To fully document the model for the reader of this chapter we still
need to describe our constraint on energy and our objective function.

The energy constraint is given in eqn. 7.3:

$$\sum_{v=0}^{t} \sum_{j=1}^{J} U_{jvtp} \geqslant q_{tp} \qquad t = 1 \ldots T, \ p = 1 \ldots P \qquad (7.3)$$

All the symbols have been defined before except:

q_{tp} = average demand in sub-period p of interval t, obtained from the load-duration curve of interval t

P = number of sub-periods in each interval.

The objective function is to minimise

$$\sum_{v=1}^{T} \sum_{j=1}^{J} c_{jv}x_{jv} + \sum_{t=1}^{T} \sum_{v=0}^{t} \sum_{j=1}^{J} \sum_{p=1}^{P} f_{jvtp} \ U_{jvtp} \ \theta_p \qquad (7.4)$$

All symbols have been defined before except:

c_{jv} = discounted capital and fixed charges for one plant of type j installed in interval v

f_{jvtp} = discounted cost of production of one unit of energy for U_{jvtp}

θ_p = duration of sub-period p.

7.2.4   Summing up on generation planning

The main stages in our worked example were: a first development program using existing and possible future generating plants, Tables 7.1 and 7.2, was put together possibly by intuition, possibly from previous studies, with respect to size, type, fuels, and time interval between the installation of plants. To do this some plant reserve margin was assumed. The optimum generation development program was then found by iteratively improving the first program by LP and possibly integer programming, according to a model similar to that shown in Fig. 7.1. In our example we used a fixed load-duration curve of the type in Table 7.3. Only a relatively minor change is needed to use a stochastic load-duration curve (see Appendix 7.4). However, the major change was to introduce more refinements into the reliability analysis. Analysing the program in Table 7.4 in terms of a LOLP[7] of 2.4 hours per year, Table 7.5 shows that the LOLP index varies considerably and, in the early periods particularly, is too high; later it is too low. Replacing the constraints in eqn. 7.1 with the more sophisticated constraints in eqn. 7.2, reflecting the predicted capability of each unit type,[8] produces an alternative programme, as shown in Tables 7.6, 7.7 and 7.8, which fits the desired reliability standard more closely.

7 2 5 Network limitations

As well as the generation expansion plan we must seriously consider the network expansion plan. In the case of network expansion we can again try to formulate an LP problem. Again we need an objective function and constraints, all expressed in linear form. We use the same principle as before, the operational simulation taking into account the different I^2R losses in the circuits. Power-system planners in the World Bank are presently formalising the treatment of losses as a constraint against which to optimise the system development plan in a similar manner to the reliability constraint described earlier.

With respect to the different I^2R network losses, we can approximate the squared term by a number of discrete linear sections, entered as extra linear constraints into the planning optimisation model. The network plant selection problem itself is formulated from the particular circumstances appertaining, e.g. whether dealing with high or low voltage, radial or meshed networks. We also have different kinds of overhead lines and cables. Because even the losses component can be linearised, all network effects can be read into the model as linear constraints on the objective function of eqn. 7.4 or some modification of it.

Again, we are dealing with a dynamic and not a static problem. When dealing with this in the generation planning case we defined a number of intervals (Tables 7.5 and 7.7). In the network case we again use intervals, but this time have a much greater interaction between time periods, e.g. the events of period 1 affect period 2, events in periods 1 and 2 affect period 3, and so on. It follows that decisions in the planning process must be integrated decisions over time. As before, reliability aspects enter as multiple, group and multiple-group constraints, which can create a very complex situation. Simplification is usually introduced by making semi-arbitrary rules from past experience.

The classic case, usually quoted on reliability aspects of networks, is that of substations, where an extract from a list of rules would look something like this (the numbers are arbitrary and are not taken from any actual case):
A substation must have at least 2 circuits feeding it to be considered reliable.
A group of 3 substations, chosen by multiple means, must have at least 4 circuits to be considered reliable.
A group of 5 substations, chosen by multiple means, must have at least 6 circuits to be considered reliable;
and so on.

There are ways to overcome this arbitrariness of the rules. Put simply, the number of circuits can be expressed as constraints, but it is difficult to search out a pattern of grouping. Although it might seem merely tedious, it is in fact a very difficult task, with important differences between very similar grouping patterns; it is thus difficult to draw boundary lines between multiple grouping patterns. Normally this is done manually, but some attempts have been made to computerise it. There are over 1000 constraint groupings to a simple system containing only 10 substations.

However, the total costs of the different alternative groupings will normally not vary very much, (say 20%) and a major saving will be made in computing time if we cut out some interconnections by intuition or because of likely difficult wayleaves, geography, etc.

7.3   NETWORK CONFIGURATION SYNTHESIS

7.3.1   Introduction

In electricity-supply network planning[9] it is customary to <u>first</u> put together the system configuration, usually adding to an existing network in some heuristic way, and <u>then</u> to analyse the result in accordance with the rules of power-system economics, the purpose of the latter being to ensure that the network will operate technically satisfactorily and is the least-cost solution. It is, however, possible to synthesise the network against technical and economic constraints in a much more precise and scientific way[10], as is briefly described below.

Any change of circuit impedance between any two nodes in a network directly affects the load flows in <u>all</u> branches; thus we must consider the whole system when making any configuration changes. On these grounds alone, most heuristic methods of adding to a network break down, as they rely on considering only piecemeal additions. It is, in fact, probably wiser to avoid heuristics as far as possible in that they are constrained by individual thought processes, unless the position is cut and dried in a particular instance.

In fact, we can be quite scientific about how we determine how to add to a network, using mathematical combinatorial theory although, as we shall see later, the size of the problem might well be too great for even modern computers to handle.

7.3.2   Minimum circuit length criterion

In rural networks and in some urban situations, the main criterion for determining the least-cost solution is the total cost of the copper in the distribution circuits which, to be kept to a minimum, requires the total circuit length to be kept to a minimum. Because each load point or node must have at least one circuit into it, our two main constraints are then:

$$\sum_{\substack{i=1 \\ i \neq j}}^{N} x_{ij} \geqslant 1 \text{ for all } j \tag{7.5}$$

$$\sum_{\substack{i=1 \\ i \neq j}}^{N} \sum_{\substack{j=1 \\ j \neq i}}^{N} x_{ij} = N - 1 \tag{7.6}$$

and we minimise

$$\sum_{\substack{i=1 \\ i \neq j}}^{N} \sum_{\substack{j=1 \\ j \neq i}}^{N} l_{ij} x_{ij}, \ x_{ij} \text{ to be integer,} \tag{7.7}$$

where there are N load points, with routes x_{ij} between each pair of points i and j of length l_{ij}.

We solve the above by the normal processes of linear programming (see earlier in Chapter 7). To take short cuts we can choose paths with ascending order of length, omitting all paths completing a loop, until all load points are connected.

7.3.3 Group transfer criterion

However, once we know the electrical transfers between nodes, the present capability of the network to transmit electricity, and that requirement for the future, can be precisely defined in terms of the transfers required to all possible groups of nodes. The amounts of these electricity transfers required gives us the size or scope of the circuits needed in any particular case; these circuits can be made up from connections to any node within a group of nodes to any node outside that group. We can also write down the important operational/security constraints defining the minimum numbers of possible circuits we wish to connect to each group of nodes, and subject to which constraints the cost in total of all the circuits in the network can be minimised. To enable a solution to be found mathematically we must free one node from the above process; it is usually called the 'slack node', analogous to the 'slack busbar' in load-flow analysis.

We must then:

$$\text{minimise} \quad \sum_{\substack{i=1 \\ i \neq j}}^{N} \sum_{\substack{j=1 \\ j \neq i}}^{N} x_{ij} \, C_{ij} \qquad (7.8)$$

$$\text{subject to} \quad \sum_{\substack{i \, \epsilon \, g \\ j \, \epsilon \, N-g}} x_{ij} \geqslant V_g \qquad (7.9)$$

for all possible groups g, excluding the 'slack node', and x_{ij} to be integer, where

C_{ij} = combinations of paths between nodes i and j

V_g = minimum number of possible circuits (i.e. the variables) into all groups

The rest of the notation is as before.   Fig. 7.2 illustrates the method.

As the value of N increases, both the number of variables and the number of constraints increase exponentially.   N nodes mean $^N C_2$ inter-nodal paths and, therefore, x_{ij} variables.   The total number of groups, and hence constraints for solution, will be given by:

$$^{N-1}C_1 + {}^{N-1}C_2 + \ldots \ldots {}^{N-1}C_{N-1} = 2^{N-1} - 1$$

and a modest 10 nodes requires 45 variables and 511 constraints
         20   nodes   requires   190   variables   and   524,287
         constraints.

In this pattern there is a similarity to section 7.1.1 in this Chapter, and, although there are short cuts and the 'number-crunching' capacity of computers is today large, general application of this method is still formidable.

In a practical situation many group constraints are oversatisfied within a feasible solution; thus we can first obtain a solution, using the constraints expected to be just satisfied, then substitute this back in all of the constraints.   If any are not satisfied, we repeat the solution with these included.   By this simplifying method we reduce the number of constraints and/or can tackle more nodes.   Some designs obtained by this method are illustrated in Fig. 7.3, and we can go further with making simplification assumptions in an ad hoc way; for example:

• existing circuits between nodes j and k can be included by constraints $x_{jk} \geqslant$ number of existing circuits
• for routes not feasible, considered improbable, etc., variables are either not considered at all, or the number considered is greatly reduced.

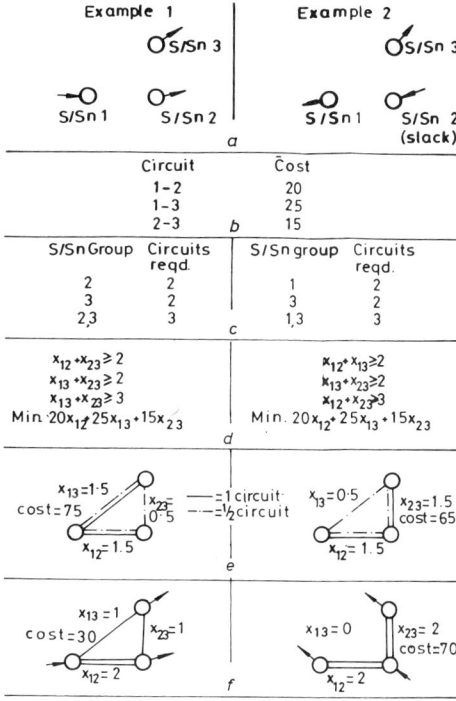

Fig. 7.2 Synthesis of network using group transfers and integer linear programming

    a   Physical positions
    b   Circuit costs
    c   Circuit requirements
    d   Integer linear programming formulations
    e   Initial non-integer solutions
    f   Final integer solutions

KNIGHT, U.G.: 'Power systems engineering and mathematics', Fig. 7.3, Pergamon Press, 1972. Reproduced by permission of Pergamon Press.

7.3.4 Perturbation methods of solution

A perturbation method (Reference[10] p. 149) can mathematically and formally be used for simplification, instead of the more practical methods described above. Still using the group-transfer criterion we can take advantage of the fact that x_{ij} in all of the above described

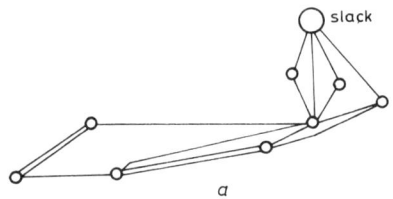

(With higher security standard than b)

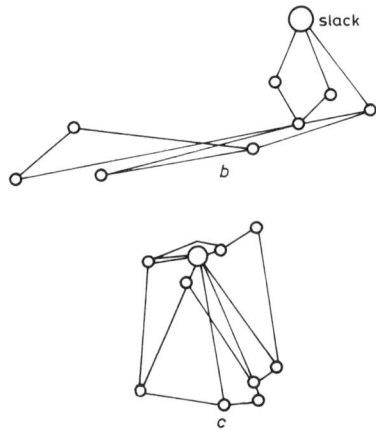

Fig. 7.3   Synthesis of network with simplifying assumptions

> KNIGHT, U.G.: 'Power systems engineering and mathe-
> matics', Fig. 7.4, Pergamon Press, 1972. Reproduced by
> permission of Pergamon Press.

analysis will probably have only values 0, 1 or 2, and not just any
integer. We prepare a technically viable solution, obtained by hand,
$[x_{ij}]_0$, and the x_{ij} are substituted in the design constraint equation,
i.e. eqn. 7.9. The $x_{ij} > 0$ with the highest cost, which appears in
an oversatisfied constraint, is decreased by 1 ($\Delta x_{ij} = -1$). If we
find that all constraints, including this variable, remain satisfied,
then:

$$(x_{ij})_1 = (x_{ij})_0 - 1$$

and this is repeated for other variable until further reductions,
which do not lead to corresponding increases elsewhere in the network,

are found no longer to be possible. Then the new solution $[x_{ij}]_1$
is substituted in <u>all</u> the group constraints, giving values $[v_g]_1$.

We can also use a decomposition method (Reference[10], p.151) for
increasing the size of the network tackled. We proceed as follows:
we assume some feasible solution is available. We choose any subset
of the nodes in this feasible solution, and assume that circuits
between this subset and the rest of the network remain unchanged;
each node in the subset can be classified as having connections
either: (i) purely internal to the subset, or (ii) internal and
external connections. For each of type (ii) nodes, by scanning all
connected groups of nodes external to the chosen subset containing
that type (ii) node, we can determine the minimum connection necessary
between it and the remainder of the nodes in the subset.

Hence, in so far as the chosen subset is concerned, we have replaced
the remainder of the network by circuit requirements at those nodes
in the chosen subset which have external connections. We know the
circuit requirements for all nodes totally within its chosen subset;
hence we can change the connections within the subset without
violating any security constraints, the charge being towards an
optimal solution within the chosen subset, using normal optimisation
techniques (see earlier in Chapter 7). By successive choice of the
subsets of nodes, we optimise different overlapping parts of the
network and thus arrive at a composite whole.

7.3.5 Summary

The normal procedure in network planning is still to put together
heuristically a number of possible designs for the future development
of the network, to test that these are technically acceptable, and
then to find the least-cost solution by economic analysis.
Theoretically it should be possible to proceed the other way round,
i.e. to synthesise a network configuration according to some rules
of technical and economic optimisation. However, this type of
approach has not been greatly used, mainly in the past because of
the very large number of variables and constraints involved. Although
modern computers can deal quickly and efficiently with vast matrices,
for our purposes in power-system planning, this fact is unlikely to
be sufficient to overcome the difficulties with network synthesis
in the foreseeable future.

7.4   APPENDIX: PRINCIPLES OF POWER-SYSTEM ECONOMIC MODELLING

7.4.1   Introduction

There are so many types of model in use today for power-system
economics that the choice of model rests as much with the taste of
the planner, computing facilities, data and size of the system, as
on the superiority of a method. Three groups of models, developed

early in system economic planning,[11] continue to be important:

(i)   Models to carry out marginal analysis;
(ii)   Models to carry out simulation of power system operation;
(iii)   Global models.

7.4.2 Marginal Analysis

When carrying out marginal analysis we begin by proposing a development programme of generating and network plant for a given time period ahead, usually to meet a given load at a given reliability standard, although this is now changing (Chapter 3). This first programme may be the outcome from a general review of the last power-system-planning studies carried out; or put together by intuition. Using a discount rate[12] equal to the opportunity cost of capital, we calculate the present value of the total yearly capital, maintenance, fuel, operating, and other costs attributable to the programme over its lifetime and sum them to a total present cost. A comparatively small or 'marginal' change is then applied, e.g. we might vary amounts of different types of generating plant in the programme. We find the total present cost of the development programme after its marginal change and compare it with the total present cost of the initial programme. If the total present cost is reduced by making the marginal change, then the programme with the marginal change is preferred. This process can be repeated for a variety of marginal changes until an optimum programme is reached (Chapter 8).

To make a model we need to state the marginal analysis process algebraically. Let the capital and other fixed costs of the initial programme be K_1 and those of the programme after the marginal change be K_2. Let K_1 and K_2 all occur in the same year. Let the corresponding running, i.e. operating, maintenance and fuel, costs for the two programmes in any time be U_{1t}. Then:

$$\text{Difference in PV of cost} = (K_1 - K_2) + \sum_{t=0}^{T} (U_{1t} - U_{2t})(1 + r)^{-t} \quad (7.10)$$

If this expression is positive then the marginally altered programme is preferable. The marginal changes to the initial programme may be of many kinds; we can change the mix of generating plant.by fuel, or advance/delay in time one or more generating plants, increasing/decreasing their discounted capital costs. Marginal analysis can be described as the general approach to determining a least-cost development programme by successive approximation.

At an early stage two difficulties arose with models for marginal analysis.

First, it was complicated and tedious to calculate by hand the running costs for a large, multi-generation system over a 20 to 30 year planning horizon, especially with demand continuously changing, and the system size doubling every 5 to 10 years.

Secondly, the number of credible marginal changes possible was often very large.
To overcome the first difficulty the simulation models, already used for system operation, were extended for use in system planning. To overcome the second difficulty and to help decide which marginal changes should be considered, global models were developed.

7.4.3 Simulation models

The most common form of simulation model uses a computer sub-routine to assemble the generating plant for operation in 'merit-order' of ascending marginal running costs per kWh (Fig. 7.4).

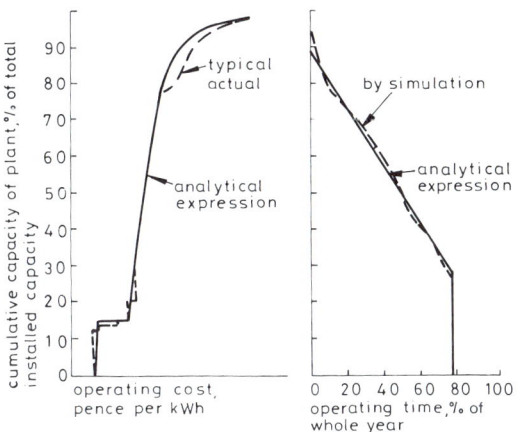

Fig. 7.4 Generating plant capacity/operating cost curve and generating plant capacity/operating time curve

BERRIE, T.W.: 'The economics of system planning in bulk electricity supply'. Elect. Rev., 22 Sept. 1967. Reproduced by permission of Electrical Review

The model then determines the optimally-used capacities of these generating plants by using a load-duration curve, also shown in Fig. 7.4. It estimates the areas optimally taken up by merit-order running within the load-duration curve for each generating plant in order to calculate the optimal electrical energy generated by each plant throughout the life of the development programme. In practice the total system running cost is found using the computer model for sample days in sample years, predicting the other days and years by interpolation.

This type of simulation model has been used in the UK for many years[13] and is similar to that used by a World Bank consultant to evaluate projects on mixed thermal/hydro systems[14], the main difference being that a special routine is required in the latter model to locate the hydro plant correctly in the load-dispatching schedule against the load-duration curve (see later).

Many possible refinements can be made to the simulation model. The UK programme mentioned above uses a linear program (LP) to determine the 'merit-order' as a sub-routine for determining the optimum pattern of coal transport between coal mines and power stations. It can also use a different load-duration curve for each season to allow for the effects of plant-maintenance schedules on system operation. An important refinement is to determine an optimum storage policy for hydro plant on a mixed hydro/thermal system. To locate the hydro plant correctly in merit-order on the load-duration curve for a particular month, we must determine in advance how much hydro energy is to be discharged in that month and how much to be stored for the next and successive months. Because this is a standard 'stock-control' problem, many engineers have used dynamic programming to find solutions.[15] Swedish engineers in particular seem to have put this method into extensive use at an early date.[16] It is also possible to use linear and non-linear programming in a simulation model to compute optimum generating plant operating schedules and costs. These can be used to solve the hydro storage problem and have proved a powerful technique to take into account transmission losses and fossil fuel transport policy; they should also be effective in dealing with nuclear-fuel cycling costs on a system with thermal and breeder reactors.

7.4.4 Global models

The second difficulty connected with marginal analysis is the large number of marginal changes we could make to an initial development programme. If we could limit ourselves to marginal changes made to plant to be installed in the immediate future, the number of changes should not be great. Unfortunately we must take some account of marginal changes in future programmes because future operating schedules and system running costs for <u>all</u> plant will depend upon the future generation composition of the system. Thus, in order to make the best forecast of generating-plant operating schedules and system running costs when analysing the effect of making marginal changes in a proposed immediate programme, we must make the best forecast we can of the optimum future generating plant programme. In this and related matters global models have proved useful. They are designed to:

(i) scan and cost a large number of present, immediate plus future development programmes; and then
(ii) select the optimum overall pattern of development programmes from the investment costs of each development programme and the total system operating costs.

The models must therefore simulate the operation of the system and calculate optimum operating generating plant schedules and costs for each relevant day/year over the period ahead for which the model is being run. Simulation models are thus not in any way superseded by global models; the models are complementary, because to arrive at a uniquely optimum development programme in one computer run by means of a global model is asking too much, even on the smallest system.

Because of their complexity, formulating global models entails approximations; however, once approximate optimal global solutions have been reached, they will be examined in more detail by marginal analysis using simulation models. As Bessiere and Petcu have argued, global models and marginal analysis are also complementary.[17]

By way of illustration we now discuss the basic global models developed in the 1960s in that they are illustrative of models now in use. One employs linear programming (LP), the other non-linear programming (NLP). LP was first studied for use on the Electricité de France power system in the mid-1950s by Masse and Gibrat, who published their results in 1957.[18] This may have been the first application of LP to power-system planning. Since then LP has been used extensively in many countries[19], and extended to cover the analysis of investments in the entire energy sector.[20]

The LP philosophy now described is that developed in the World Bank, see (Reference 11). This is a good model to illustrate LP as it follows previous modelling practice fairly meticulously. It has been applied with success, especially to the hydro-thermal system in Turkey.[21]

The load-duration curve is broken down into $p = 1....P$ discrete blocks, see Fig. 7.5. The unknown MW capacity of plant j installed in year v (vintage v) is denoted by X_{jv} and the discounted capital cost is C_{jv}. The unknown MW output of plant jv in period p of any subsequent year t is denoted by U_{jytp}, and its operating cost per MWh in that period by F_{jyv}, the width of the period being θ hours.

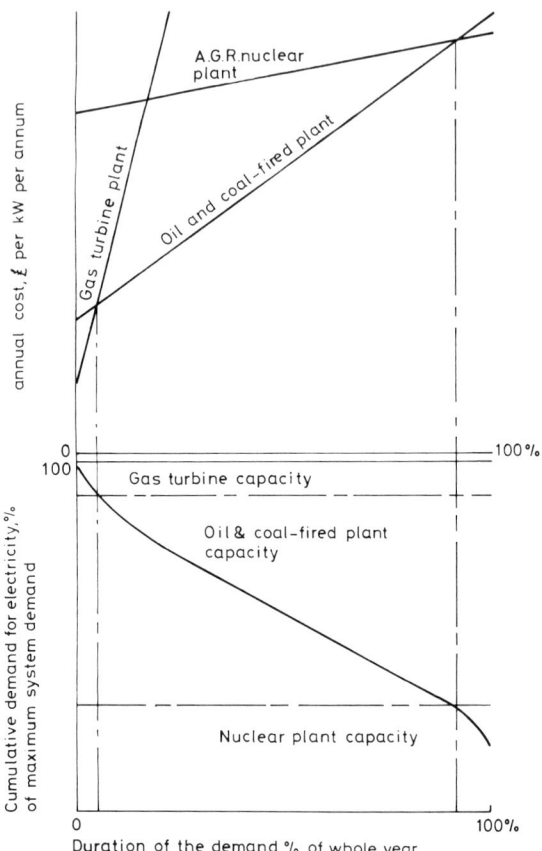

Fig. 7.5 Determining the plant mix

BERRIE, T.W.: 'The economics of system planning in bulk electricity supply'. <u>Elect. Rev.</u>, 22 Sept. 1967. Reproduced by permission of the Electrical Review. For details of the method see Electrical Review article

The objective is to choose plant capacities and plant outputs so as to:

$$\text{minimise} \quad \sum_{j=1}^{J} \sum_{v=1}^{T} c_{jv} x_{jv} + \sum_{j=1}^{J} \sum_{t=1}^{T} \sum_{v=-V}^{t} \sum_{p=1}^{P} f_{jtv} u_{jtvp} \theta_p \qquad (7.11)$$

where T denotes the period of the study, usually 30 years or so, and $v, -V \ldots 0$ denotes the vintages of plant on the system at time $t = 0$.

The principal constraints are that aggregate plant output must be sufficient to meet the demand in each period; the energy constraints:

$$\sum_{j=1}^{T} \sum_{v=-V}^{t} u_{jtvp} \geqslant q_{tp} \quad p = 1, \ldots, p, \quad t = 1, \ldots, T, \qquad (7.12)$$

and that the output of each plant must never be greater than the plant's available capacity; the capacity constraints:

$$u_{jvtp} \leqslant a_{jv} \cdot x_{jv}$$
$$u_{jvtp} \geqslant 0, \forall_{jvtp}$$
$$0 \leqslant x_{jv} \leqslant n_{jv}, \ x_{jv} \ \text{integer} \ \forall_{jv}$$
$$\text{for } j = 1 \ldots J, \ p = 1 \ldots P, \ t = 1 \ldots T, \ v = -V \ldots t \qquad (7.13)$$

Where a_{jv} is the plant's availability factor. There are also several other constraints stated here in words:

(i)   Available aggregate plant capacity must exceed mean expected peak demand, plus an allowance for 'reserve' in case of plant failure.

(ii) Amount of capacity of any given type is limited by the total availability of that plant.

(iii) Capacities of plant initially on the system X_{jv}, $v = (-V \ldots 0)$ are known.

(iv) Output of hydro plant is a function of water inflows and storage capacities.

(v)   Certain plant policies, especially in the short run, are pre-defined, i.e. they are not counted as decision variables.

(vi) Aggregate hydro plant capacity is limited by energy reserve requirements to some pre-defined level.

Other constraints, for example to limit capital investment intensity or foreign exchange expenditures can also be allowed for.

The above equations 7.11 - 7.13 are suitable for a standard LP program using standard software packages. The general approach can also be extended to cover many other kinds of problem:

(i)   Optimum generating plant replacement
(ii)   Optimum generating plant locations of bulk energy
(iii)   Optimum storage capacities and storage policies of hydro-electric plant
(iv)   Special constraints on the operation of multi-purpose hydro schemes
(v)   Integer variables to represent the large fixed cost component of hydro and nuclear schemes (see later).

Several other extensions are possible, e.g. constraints to represent nuclear fuel cycling can also be introduced. An interesting aspect of these formulations is when, in the equations above, we fix the plant capacities, i.e. the X's, and we then search for the optimum plant output, i.e. the U's, we have a simulation model. Now, fixing the X's gives us a lot of extra computer space for examining operating costs in more detail, especially with modern computers. Moreover, the linear program can then be decomposed into small independent LP simulation studies for each year, giving us even more computer space. We can also use bounded variables linear programming methods to cut computing times and save yet further on computer space in the simulation model. Although such devices to save computing time are not as vital in the 1980s as they were in the 1960s, by and large the above remarks still apply; in any case all power utilities do not yet have access to a large modern computing installation.

A basic difficulty with LP models is the large number of constraints encountered in any realistic formulation of a problem. To illustrate this: if we wish to examine the development of a system over a 30 year period, broken down into, say, 6 × 5 year block periods, having 4 kinds of plant, say, nuclear, hydro, fossil and gas turbine; and with the load-duration curve broken into, say, 6 blocks; then the number of constraints will be about 1000. This is still quite a large LP for many power utilities in the world to handle, although standard computer library packages are now available that can handle many more constraints, e.g. using mixed integer continuous variables facilities (see later). It is also possible, often necessary, to use matrix generators to process the very large inputs.

The principal reason for the large number of constraints is that we must make sure that the output of every generating plant on the system, in every period of the study, on every 'block' and 'period' of the load-duration curve, does not exceed its maximum capacity. Let us say there are, on average, 50 plants on the system and that we break the load-duration curve into

10 blocks, and that, as we assumed above, we take a 30 year period broken into 6 × 5 blocks, we would now have 50 × 10 × 6 = 3000 constraints of this type. This is already three times the number of constraints shown above.

In the early 1960s when this problem was first encountered, computers could not easily handle anything like this number of constraints. Possibly the reason why Electricité de France turned to NLP was to overcome this problem.[22] The UK Central Electricity Generating Board in the mid-1960s also reported using a non-linear program.[23] The (then) novel idea of this latter approach is to pre-arrange all the plant that is or may be connected to the system in any year in merit-order in the data input. This means that the operating plant sequence is decided in advance by inspecting the marginal operating costs before the computer run is commenced. By this device all the operating variables and their associated capacity constraints are satisfied implicitly and can be deleted from the algebraic formulation. The plant capacities X_1, X_2, X_3...etc. in each year are projected by means of the load-duration curve as described earlier. The areas 'sliced out' of the load-duration curve are then weighted by the marginal running cost of each plant. To get the total system running cost we integrate the weighted area under the curve. The important thing to note is that this cost can be expressed as an arbitrary, nonlinear function of the as-yet-unknown plant capacities. We then add this to the total capital cost, also expressed in terms of plant capacities, and search for the values of those capacities that will minimise the total (capital plus running) costs. The main constraint is that the aggregate capacity must be sufficient to meet the peak demand on the power system.

Many power-system planners today see no great advantage to be gained by turning to NLP. They are convinced that by using NLP we lose all the considerable advantages of flexibility and of computer software that are available with LP. LP packages are now available for every computer operator, whereas even in the early 1980's with NLP we still often have to start from scratch and write new algorithms. Again, we can quickly rewrite an LP formulation to cover new problems, e.g. generating plant replacement, bulk electrical energy transmission, multi-purpose hydro schemes etc., and still use standard computer software without having to rewrite the algorithm.

In the 1960s planners asked whether we can retain an LP form and yet reduce the constraint problem. Since that time the answer has been an increasingly confident 'yes'. It was quickly realised that the reason the NLP form reduces the number of constraints is not strictly because it is intrinsically more economic than LP, but because, in the NLP model, we include a priori information about system operating characteristics which we exclude from LP models. It is only because this information is definitely excluded from the LP models that we get so many constraints in the LP models. If we could include it then we can reduce the problem size to virtually the same numerical proportions as we encounter with NLP. The way this information was first included was as follows.[24] We know that, as we move along the time axis of the 'load-duration' curve, the output of any plant will not be increased. It will either remain the same as before or will be reduced. Suppose that we define new operating decision variables, Z's, to replace the U's, which until now have represented the output of each plant. These Z's are defined to be the <u>decreases</u> in output of any plant as we move along the load-duration curve. For any plant j:

$$z_{jp} = U_{jp} - U_{jp+1} \geqslant 0 \tag{7.14}$$

We also know that the total decrease in output as we move along the load-duration curve can never exceed the plant capacity, so that:

$$\sum_{p=1}^{P} z_{jp} \leqslant x_j \quad j = 1 \ldots J \tag{7.15}$$

We can thus replace the U's, which must satisfy plant capacity constraints on every portion of the load-duration curve, with new non-negative variables, the Z's, which satisfy one constraint for the whole curve.[25] If P is the number of blocks into which the load duration curve is divided, the number of constraints is reduced by 1/P. For example, our 1000 constraint problem above is now reduced to about 170 constraints; our 3000 constraint problem above to 300 constraints. A 10000 constraint problem might be reduced to 600 constraints by 'improved-Z-variable' methods which have more recently been developed. This restores LP to its former versatility, i.e. without resorting to NLP.

7.4.5   Combined cost method

A dramatic change can be made in the number of constraints by the combined costs method (CCM)[26], which removes operating constraints completely from the global model described above. By this means the number of constraints is reduced by up to a hundredth of that required in the early LP models.

In CCM values are assumed for the plant factors for non-energy limited (NEL) plants, i.e. fossil fueled and nuclear plants. Constraints are imposed only on energy limited (EL) plants, e.g. hydro and geothermal. For NEL plants the actual and the anticipated plant factors are one and the same; this provides for a large degree of simplification.

Besides eqns. 7.11 - 7.13, we now have the equation for all
EL plants:

$$f_{jt} x_{jt} h_t = e_{jt} \qquad (7.16)$$

where f_{jt} = plant factors associated with X_{jt}
       h_t = annual load-duration in hours in year t
       e_{jt} = maximum energy available for generation using energy
               limited plants in year t

The addition of this equation to the LP model ensures the full
exploitation of energy from EL plants and there are some minor changes
to the previous equations given in this Appendix.

The problem written in CMM form can be solved as an LP model to
get the optimal generation capacities as before.

7.4.6 Integer programming

It has been realised for some time that the discrete nature of
generating unit sizes of a particular plant type requires an integer
solution.[27] Hence the normal LP algorithms are not strictly suitable
and the discrete nature of the problem has to be treated by integer
programming (IP). The IP solution depends on assumptions about the
initial plant factors; this requires simulation of the proposed power
system to obtain the expected operating costs and consequently the
expected plant factors. These expected plant factors will be
different from those assumed in the original IP formulation, forming
a better set of values to be used for the next IP solution; this
iterative nature makes the IP method very attractive to the
practitioner since he can intervene between successive solutions.

An up-to-date programming suite would also probably contain a
'branch and bound' technique for optimisation, i.e. of moving forward
through successive steps in time in the optimum way.[28] The static
load-duration curve with which we have been dealing to date would
also probably be replaced by a probabilistic load-duration curve.[29]

REFERENCES

[1]   LAUGHTON, M.A.: 'Electric power systems - models for expansion planning'. Paper given at International Symposium on Electricity Economics and Load Management, Imperial College, London, 24-28 March 1980. A good deal of this chapter is based on this reference by permission of Power Systems Group, Imperial College

[2]   BERRIE, T.W.: 'Further experience with simulation models in system planning'. Second Power Systems Computation Conference, Stockholm, June 1966; also BERRIE, T.W.: 'The economics of system planning in bulk electricity supply', in, TURVEY, R.(Ed.): 'Public enterprise' (Penguin 1968); also JONAS, P.J.: 'A computer model to determine the economic performance characteristics of the British generating stations'. British Computer Conference, Brighton, 1966

[3]   TURVEY, R., and ANDERSON, D.: 'Electricity economics, Chapter 5'. Published for the World Bank (Johns Hopkins University Press, Baltimore, USA, 1977)

[4]   SULLIVAN, R.L.: 'Power system planning' (McGraw-Hill, 1977)

[5]   NORRIS, T.: 'Economic comparisons in planning for electricity supply'. Proc. IEE, 1970, 117, (3), pp.593-605

[6]   ADAMS, R.N., and LAUGHTON, M.A.: 'Optimal planning of power networks using mixed-integer programming', Proc. IEE, 1974, 121, (2)

[7]   SULLIVAN, R.L.: 'Power system planning'. (McGraw-Hill, 1977) p. 221

[8]   'The prediction of reserve requirements in generation planning', Int. J. of Elect. Power and Energy Systems, 1980, 2, (1)

[9]   BERRIE, T.W.: 'Economics of system planning in bulk electricity supply' in TURVEY, R.(Ed.): 'Public enterprise' (Penguin Economics Classics, 1967) Chap.6

[10] For a much fuller explanation on network synthesis the reader is referred to KNIGHT, U.G.: 'Power systems engineering and mathematics' (Pergamon Press, 1972)

[10^a] Reference [10] page 149

[10ᵇ] Reference [10] page 151

[11] For a fuller description of the main three types of models mentioned here and their history see: BERRIE, T.W., and ANDERSON, D.: 'Power system planning, development programs and project selection – a discussion of methods'. Paper given at Fourth Power System Computation Conference, Grenoble, France, 11-16 Sept. 1972. The author uses some of his earlier arguments and the notation of this paper in this Appendix

[12] GETTINGER, J.P.: 'Compounding and discounting tables for project evaluation'. Published on behalf of the World Bank (Johns Hopkins University Press, Baltimore, USA, 1973)

[13] see Reference 2

[14] JACOBY, H.P.: 'Analysis of investments in electric power'. Center for International Affairs, Harvard University, 1967

[15] LITTLE, J.D.C.: 'The use of storage water in a hydroelectric system'. J. OR Soc. Am., May 1955, (2); also MASSE, P.: 'Les réserves et regulations de l'avenir dans la vie économique' (Hermoun, Paris, 1946)

[16] LINDQVIST, J.: 'Operation of a hydrothermal electric system; a multi-stage decision process', IEE Trans., April 1962

[17] BESSIERE, F., and PETCU, M.: 'Analysis marginale et optimisation structurelle des investissements: Application au secteur de l'electricité', RIRO, Paris, 1967, (6), p. 61

[18] MASSE, P., and GIBRAT, R.: 'Application of linear programming to investments in the electric power industry'. Reprinted by NELSON, R.J.(Ed.): 'Marginal cost pricing in practice' (Prentice Hall, 1964)

[19] 'Symposium on the application of OR methods in the solving of economic planning and operations of large electric systems and on the use of computers for that purpose'. United Nations Economic Commission for Europe, Varna, Bulgaria, May 1970. Also see Proceedings of First, Second and Third Power Systems Computation Conferences, London, 1962, Stockholm, 1966, and Rome, 1969

[20] FORSTER, C.I.K., and WHITTING, I.J.: 'An integrated mathematical model of the fuel economy', Statistical News London, Nov. 1968, (3), to name an early example which is still probably the clearest to understand; also ANDERSON, D., and TARKAN, O.: 'Optimum development of the electric power sector in Turkey'. IBRD Economics Department Working Paper, No. 126, March 1972; this latter model also examined the overall energy sector

[21] Economics Department Working Paper 91, op.cit.

[22] BESSIERE, F.: 'Methods of choosing equipment at Electricité de France: Development and present-day concept', European Econom. Rev., Winter 1969, 1, (2)

[23] JENKIN, F.P., PHILLIPS, D., PRITCHARD, J.A.T., and RYBICKI, K.: 'A mathematical model for determining generating plant mix'. Third Power Systems Computation Conference, Rome, 1969, vol. 2

[24] The idea is due to BEALE, E.M.L., and WHITTING, I. The author is indebted for many an early discussion on these themes with the latter

[25] ADAMS, N. et al.: 'Mathematical programming systems in electrical power generation and distribution planning'. Fourth Power Systems Computation Conference, Grenoble, 1972

[26] BEGLARI, F., and LAUGHTON, M.A.: 'The combined cost method for optimal economic planning of an electric power system'. IEEE Trans., 1975, PAS94, (6), pp. 1935-1941

[27] For a much fuller description see FERNANDO, P.N., INDURUWA, A.S., CORY, B.J. and MCKECHNIE, A.: 'Further developments in generation planning using integer programming'. Proceedings of Sixth Power Systems Computation Conference, Darmstadt, Aug. 1978, pp. 12-21

[28] FERNANDO, P.N.: 'The planning of electical power system generation and transmission using integer programming and reliability concepts'. Ph.D. Thesis, University of London, 1976

[29] SULLIVAN, R.L.: 'Power system planning' (McGraw-Hill, 1967); also BILLINGTON, R.: 'Power system reliability evaluation' (Gordon & Breach, 1970)

Chapter 8

Alternative generating projects

8.1 CHOICE OF METHODS

8.1.1 Introduction

Generating projects are usually selected within the context of a previously determined background plan as described in Chapter 5. In the early years the plan may contain information on particular stations, especially when power utilities roll their development plans forward every year or two. Using the background plan, marginal system running costs are determined, theoretically, for every hour during the period covered by the background; such information is usually provided by a computer simulation model (see Chapter 7). There are various ways of using these marginal system running costs, and the capital costs of individual generating plant to arrive at an economic ranking of these plants.[1] All costs should be in real terms (i.e. containing only relative inflation) and be expressed in economic terms; i.e. some 'shadow' pricing may be necessary.

To be ranked the projects must first be fully identified, i.e. sited, typed by fuel(s) and timed so that their capital and running costs can be determined with sufficient accuracy. The utility must continually review its sector for growth, transmission routes, availability and prices of fuels, plant sites, construction times and capital costs. It is important to consider a wide variety and number of different alternative plants; five to ten sites should be considered for each economic ranking exercise depending upon system size. Some important questions are:

(i) Will the total present value of the system costs be reduced by the introduction of the project? This is the test of marginal analysis.

(ii) If the answer is 'yes', then how many of this type of project should be ordered?

(iii) If there is more than one favoured project, in what order should the individual plants be installed; i.e. there may be a preferred 'sequence' of plant installation (see Appendix 8.5).

(iv) How will the likely relationship of capital to running costs

of plant to be installed in the future affect the decisions
on investment now?

8.1.2   Constant-load-factor method

One way of answering the questions is to compare the total cost
per kWh of generating electricity at constant load factor by different
alternative generating plants. The capital investments for different
projects are converted into an annual charge, i.e. an 'annuity',
taking proper account of their economic lives and appropriate rate
of interest on capital.[2] The annual charges for the alternatives are
then divided by the total kWh generated throughout the year, assumed
to be the same for each alternative and for each year throughout their
economic lives. The annual charges per kWh are added to the running
costs per kWh to give a total cost per kWh for each alternative.
The alternatives can then be ranked according to the total costs per
kWh.

The basis for providing the alternatives are plants likely to be
on offer at a future date. These are difficult to describe precisely[3]
or cost accurately because of continuing technological change;
planners must consider how far past experience in decreasing capital
cost per kW of generating plant in real terms and increasing fuel
prices in real terms can be assumed into the future. Also, some
elements in generating plant design have still to be fully tested
in practice, e.g. some nuclear and the largest conventional thermal
plant, especially when safety or environmental factors are important
considerations. This point applies especially to smaller systems
of developing countries; these countries have not operated either
nuclear or large plant, and environmental aspects are increasing in
importance in system planning. For comparing alternative plants
which can reasonably be expected to operate at the same constant load
factor throughout their lives the 'cost per kWh' method is sufficient.
An example of its presentation is given in Table 8.1.

However, specific difficulties with the 'cost per kWh' approach
are twofold:

(i)   Certain ground rules must be specified, notably the assumption
       of a constant lifetime load factor, e.g. 75% for base-load
       comparisons, 30% for mid-merit comparisons, and 10% for
       peak-load comparisons. Because the load factors of most plants
       being compared for immediate installation are often very
       different, e.g. peaking versus base-load plant, assuming a
       constant load factor is often unrealistic.

(ii)   A decision often depends on small differences (say, 5% to 10%)
       between the total generation costs of alternatives, i.e. well
       within the accuracy of the data, and often with much smaller
       differences than sh wn in Table 8.1.

Difficulty (i) can often be overcome without departing too much

Table 8.1   Estimated costs of generation at power stations at 75% life-time load factor

Type	A				B	
Station	P	Q	R	S	T	U
Station output MW	276	500	580	1,180	1,200	1,320
Capital cost per kW, cost units	185	155	106	107	81	71
Generation cost per kWh						
Capital charges	0.86	0.73	0.49	0.48	0.37	0.33
Running cost	0.37	0.30	0.21	0.17	0.15	0.15
Total generation cost	1.23	1.03	0.70	0.65	0.52	0.48

Based on: BERRIE, T.W.: 'Appraising the economic worth of alternative projects'. Elect. Rev., 29 Sept. 1967, Table III. Reproduced by permission of Electrical Review

from the 'cost per kWh' concept.[4] When comparing two generating plants running at different constant load factors throughout their lives, not too much distortion is obtained by calculating the total cost per kWh at a third constant load factor which is chosen by experience intermediate between the two other load factors. When comparing two plants which have identical load factors but which vary over time, the minimum distortion is introduced by calculating the total costs per kWh at a load factor represented by the 'present value' of the common load factor.[5] Difficulty (ii) is a problem in all methods of comparison. The advice given in Chapter 4 about significant figures very much applies.

8.1.3   Net effective cost method

   In Chapter 5 we described in detail why there will be system fuel cost savings made by introducing most new plant which can be measured over its economic life. From Chapter 7 we know that these fuel cost savings are usually calculated for 'typical' days in 'typical' years throughout the economic life of the plant, the savings for other years being determined by interpolation. Bringing each annual fuel cost saving to a 'present value' we calculate the total present value of system fuel cost savings for a particular generating plant over the economic life of that plant.
   The net effective cost of the plant is found by subtracting the total present value of the system fuel cost savings from the capital cost, the latter including capitalised interest during construction, and the cost of associated transmission and ultimate scrapping. All costs

Table 8.2   Basic estimates of net effective cost of future stations
            (March 1980 price levels)

		Nuclear £/kWpa	Coal-fired £/kWpa
A	Capital charges at station and provision for decommissioning, interest during construction	77	36
B	Inclusive fuel costs	34	113
C	Other costs of operation	12	10
	Generating costs	123	159
D	<u>Less</u> fuel savings from displacing less efficient plant	148	143
	Net effective cost excluding transmission*	−25	+16

* In each case this amounts to £5/kW per annum for capital charges
and interest during construction together

Source: 'Central Electricity Generating Board'. Monopolies and
Mergers Commission Report, Table 5.2, HC315, HMSO, London, 20 May
1981. Reproduced by permission of HMSO. Original source: Central
Electricity Generating Board

are usually expressed in the form of an annuity taken over the life
of the plant (Table 8.2).

If the net effective cost is negative, then not only should that
particular generating plant be selected, but often that plant with
those capital to running costs should continue to be added until the
net effective cost turns positive, regardless of system growth. For
some systems today, possible negative net effective costs are
characteristic of generating plants with low or comparatively low
fuel costs, e.g. hydro or nuclear plant, even though these plants
have high capital costs. These generating plants have a negative
net effective cost because of the very high price of fuels presently
being used by the bulk of the existing power stations on the system.

Power systems are ranked according to the size of the negative net
effective cost; i.e. the smallness of their net effective cost if
this is positive.

In calculating the net effective cost only associated transmission
is included, i.e. that necessary to connect the plant to the
interconnected system. It is not easy to take detailed transmission
costs into account. This is one reason why the method is at its best
when applied to the large interconnected systems found in Europe and

Japan. We deal with geography in Chapter 10.

A project with a positive net effective cost may still be cost effective regardless of system growth, because that project may be less costly than the alternative of retaining existing plant. The relevant figure for comparison is the 'net avoidable cost', i.e. the cost of keeping plant available to take up load at short notice, including wages, salaries, rents, rates, insurance, repairs and maintenance, and an allowance for overheads. The net avoidable cost of deferring retirement is one yardstick against which potential cost-effective new investment is measured. It is therefore expressed in the same units as the 'net effective cost', e.g. £/kWpa.[6]

In the case of nuclear plant it is important to define fairly precisely the fuel costs. A breakdown favoured in the UK is to divide the nuclear-fuel cycle into the following elements:

(i)   Procurement of natural uranium
(ii)   Enrichment of the fissionable isotopes
(iii) Fabrication of fuel elements
(iv)   Management of irradiated fuel.

In the UK each item accounts for about a quarter of the estimated present value of the nuclear-fuel-cycle costs used in calculating the net effective cost of Table 8.2.

As with all methods, it is important to test the net effective costs against changes in the main parameters. Table 8.3, taken from a different base than Table 8.2, gives an example of this.

Table 8.3   Net effective cost of new stations (March 1980 prices)

	Nuclear plant £/kW pa	Coal-fired plant £/kW pa
Basic net effective cost	−23	+21
Net effective cost if:		
Nuclear construction cost + 15%	−10	+21
Nuclear construction cost − 15%	−36	+21
Coal-fired construction cost + 15%	−23	+27
Coal-fired construction cost − 15%	−23	+15
Coal prices increase of only half that assumed in the base case	+9	+27

Source: 'Central Electricity Generating Board'. Monopolies and Mergers Commission Report, Table 5.12, HC 315, HMSO, London, 20 May 1981. Reproduced by permission of HMSO. Original source: CEGB Memorandum M17 to the House of Commons Select Committee on Energy

8.1.4   Applicability of the net effective cost method

   The net effective cost method is difficult to apply for small
systems, and for the smallest virtually impossible because the method
is intensely incremental, i.e. it examines very small incremental
changes in the total fuel cost of a system caused by introducing a
particular generating plant. The incremental plant capacity introduced is small (1% to 5% say) compared with the size of the total
system capacity. The method assumes that introducing that particular
plant will not materially change the background plan, e.g. from which
the basic fuel cost savings data is obtained. On smaller systems
this cannot be assumed; the background plan will change fundamentally
if one particular plant is introduced. Marginal analysis to justify
a particular generating plant must then be carried out by going back
to the background plan itself; i.e. by changing the sequences of the
different plants being installed until there is no appreciable
reduction in the net present value of costs.
   Some argue[7] that carrying out marginal analysis by the net effective
cost method and the plant sequence method is the same, once the effects
of 'indivisibilities', i.e. bulky investments, are allowed for.
However, adding a large generating plant to a small system has a
massive effect, whereas adding the same plant to a large system will
be insignificant. At some point, therefore, lumpiness becomes
crucial and changing the background plan to a marked extent becomes
necessary.
   The following points concerning the method will help planners to
decide whether the method is applicable for their system:

(i)   It was developed for a very large interconnected system
       composed almost entirely of thermal plant. We cannot rush into
       adapting the method for any other type of system, e.g. small
       thermal or mixed hydro-thermal. In the case of a hydro-thermal
       system it means first removing the hydro plant from the
       calculations by assuming that they operate according to
       exogenous rules, e.g. to optimise the use of their stored water
       throughout the periods of the year when thermal generating
       plants' fuel costs are at their maximum. This has been done
       with some success,[8] but the general applicability of the method
       to any but large thermal power systems is unlikely.

(ii)   The method was developed against a backdrop of availability
       of background plans.[9] Therefore, in reality the method goes
       somewhat beyond the idea of calculating the net effective cost
       against one background plan. This makes it more applicable
       to smaller systems than item (i) above implies, but it invites
       questions on sensitivity analysis.[10]

(iii) The method was devised for a system more or less in an
       'equilibrium of growth'. We have used this phrase already in
       a similar context in Chapter 4 for trend load forecasts. More
       work is needed to establish whether it would be a real problem

to use the method on systems with unsteady growth, or whether there is merely a problem of presentation of the results.

(iv)   The method was devised for a system whose background plan already basically contained all the alternative projects under consideration. In this sense the method does not compare mutually exclusive generating plants. The question is: does it really matter? This point can get very complicated. For example, some say the alternatives being considered are mutually exclusive because the background plan contains them only in 'outline', whereas the marginal analysis of the net effective cost method includes fairly exact parameters about these alternatives; e.g. concerning their capital costs, associated transmission costs, thermal efficiencies etc. These questions are important when using the method for small electricity systems in a state of rapid growth. In the latter case, the background plan for the early years <u>must</u> contain a large number of sequences of alternative but <u>not</u> mutually exclusive generating plants.

(v)   Much early thinking on the method centred round:
(a)   Assuming the system is one single generator.
(b)   Assuming this one single generator does not change much in its size, characteristics, fuel mix, operating regime, from one year to the next successive year throughout the life of the background plan. This may make the same point as (i) above, but may raise deeper issues and more work is needed.

8.2   SMALLER, OR RAPIDLY EXPANDING, OR MIXED HYDRO-THERMAL SYSTEMS

8.2.1   Introduction

We may well wish to use a different approach from the net effective cost method for the smaller, or expanding, or mixed hydro-thermal systems, e.g. those met with in developing countries. We go back to changing the background plan for each comparison between alternative projects. Basically in practice we must keep the first (say) ten to fifteen years of a background plan flexible and from then onwards keep the plan roughly constant, i.e. only in the form of a 'plant mix'. Because the early part of the plan is fluid we can carry out permutations and combinations of alternative sequences of installation of generating plants in those early years.

8.2.2   Comparing alternative sequences of generating plants

We thus compare directly different background plans built up from alternative sequences of generating plants which in turn include the individual generating plants we wish to compare. Fig. 8.1 illustrates this point.

In Fig. 8.1 there are three different sequences of generating plants to meet the forecast load at the optimum standard of

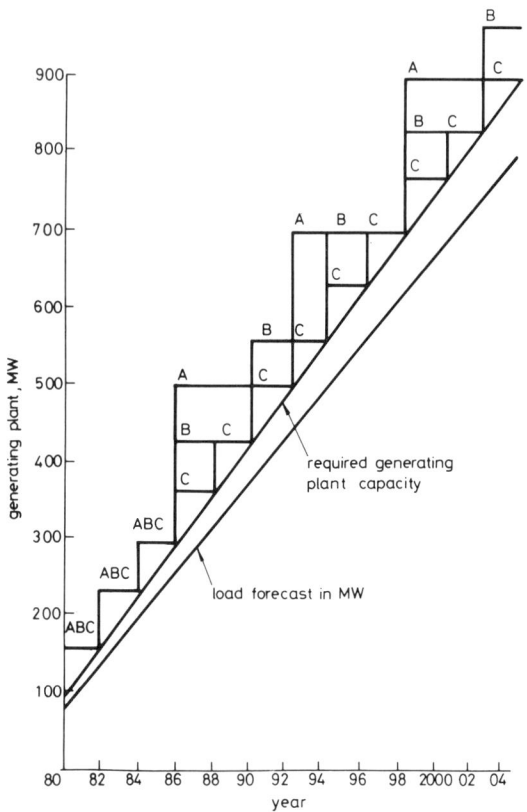

Fig. 8.1   Sequences of different generating plant developments

Sequence A : Three diesels then large steam plant
Sequence B : Three diesels then combined cycle plant
Sequence C : All diesels

Source: The author

reliability, shown as an optimum planning plant margin of 15% to 20%.
Sequences A, B and C have different capital and running costs between
1986 and the year (say) 2005, after which they are assumed to have
only differences in running costs.

Because each sequence contains plant which has an economic life
longer than the year 2005, the economic calculations for the sequences
are continued for a further (say) twenty years, using the running
costs for each sequence held constant at those for the year 2005,
until there is no difference between the present value of these
running costs, say up to 2025, the end of the period of evaluation.
Appendix 8.5 gives the detailed methodology.

8.2.3 Total present costs

The economic analysis of the different sequences of development is made by comparing the summated present values of annual capital plus system running costs of each alternative plant sequence up to the end of the evaluation period, working in constant price levels but including relative inflation.   Sequences are then ranked according to the total present value of cost, starting with the smallest.   Scrapping of plant is taken care of by including in the annual cost streams the attributable costs of keeping all generating plants operational.   In this sense shutting down an existing plant is just as much an option for including in a sequence as is installing a new generating plant. As with the net effective cost method, costs included for each plant sequence are:

(i)   The capital cost of all new generating plant, plus associated transmission in the initial period (in our case up to 2005). In contrast to the net effective cost method these capital costs are rarely expressed as annuities. Interest during construction is allowed for in the discounting.   The annual capital costs are present-valued and totalled.

(ii)   The running cost of the whole system with that particular sequence of generating plant installed.   If the sequence includes shutting down generating plant the attributable savings are included.   The annual running costs are present-valued and totalled over the whole evaluation period (in our case up to 2025).

(iii) The attributable overheads, present-valued and totalled for each sequence over the whole evaluation period.

Fig. 8.2 shows a typical presentation of the results of the economic comparison between three alternative sequences of development. Normally we trade-off the present value of lower running costs throughout the life of the generating plant in one sequence against the present value of dearer capital costs of that sequence, when compared with another sequence which has higher running costs. The point at which the total present costs of two such sequences cross is often called the 'equalising discount rate'. If the comparison is indeed the trade-off as described above we can be reasonably certain that the equalising discount rate gives some measure of the return in real terms on the additional capital cost of one sequence compared with the other. However, the most important criterion is to choose the sequence with the lowest total present value of cost at the discount rate(s) considered appropriate.

8.3 ECONOMIC RETURN

8.3.1 Introduction

It is difficult to answer the question: 'Which alternative project

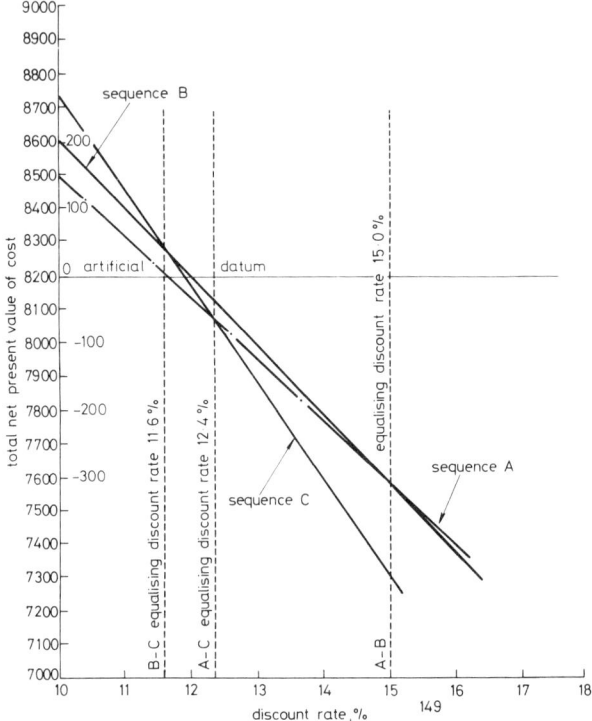

Fig. 8.2 Economic comparison of three sequences of generating plant
developments

Source: The author

is the most economic?' without inviting the question: 'Should we
invest in any project at all?' Nothing in this Chapter to date queries
whether we should invest in anything at all.

In order to address this option we must calculate the economic
return from investing in the chosen project or sequence. To do this
we need to know the economic benefits attributable to the
project/sequence.   As mentioned in Chapter 2, in the power sector
both costs and benefits in economic analysis are normally measured
with respect to the economy rather than the utility or the consumer.

The economic costs will have been worked out in great detail to
find the most favoured project/sequence.   A first measure of the
benefits to the economy are given by what consumers are willing to
pay for the electricity made available by the project/sequence.[11]

Using this as the 'benefits' of the project/sequence, the economic
return is the discount rate which makes the present value of the
attributable costs equal to the present value of the attributable
benefits over the economic life of the project/sequence. Willingness
to pay is normally measured at existing tariff levels.

8.3.2 Attributable benefits

We must say more here about attributable benefits:

Firstly: Attributable benefits are kept constant for the whole of
           the economic life of the project/sequence if the net
           effective cost method is used. Attributable benefits are
           usually kept constant over the years between the end of
           the initial period and the end of the evaluation period
           if the alternative sequence method is used. Within the
           initial period the attributable benefits are built up from
           zero at just before commissioning of the first project
           in the sequence, to a maximum at the end of the initial
           period, more or less in line with the growth of the system
           demand in kW and kWh.

Secondly: Attributable benefits are not just those measured through
           willingness to pay at the existing tariff level, although
           the increased benefits to the nation may be very difficult
           to quantify.[12] In addition to revenue, secondary benefits
           to the economy include making electricity available for
           growth; resource cost savings in fuel, e.g. by
           substituting electricity for less efficient energy
           sources such as electric-lights/stoves for kerosene-
           lights/stoves.

Thirdly: In calculating the economic return, all costs and benefits
           must be in economic and not money terms.[13]

Economic returns must be subject to a sensitivity analysis of the
main parameters, e.g. fuel cost, capital cost, foreign exchange rate,
load forecast and tariff level.

8.3.3 Acceptability of the economic return

Many governments and lending agencies lay down an acceptable level
for an economic return on an investment programme in the public
sector. Because the return is calculated in real terms, i.e. not
allowing for general inflation, but only for relative price
inflation, a lower figure for the economic return is often sought
than for the financial return on private capital. A figure of 3%
to 5% is acceptable in real terms in some developed countries, but
a much higher figure of 8% to 12% is required in some developing

countries where capital is very scarce in an attempt to achieve some measure of capital rationing.

8.4 SUMMARY

The scope and depth of the method used to choose between alternative generating projects or development programmes should be appropriate to the circumstances. When comparing different projects with very similar outputs, these can be ranked according to increasing total (capital plus running) cost per kWh generated. In all other cases it is necessary to write down a cash flow of attributable capital and running costs over the economic life of the alternative projects/development programmes, bring these cash flows to a total present value of cost (TPC), and to rank the projects/programmes according to ascending value of TPC. It is necessary to test all ranking against a variation in the main parameters by means of a sensitivity analysis.

Choosing between alternative projects is not the same as deciding on whether any project should be gone ahead with. The test for the latter is an acceptably high economic return on the original investment in a project/development programme, including the attributable benefits (mainly revenues) over the economic life.

8.5 APPENDIX: COMPARISON OF ALTERNATIVE PROJECTS

8.5.1 Introduction

In our treatment of the economic comparison[14] of projects in this book we assume that costs and benefits are measured with respect to the national economy, not to the power utility or the electricity consumer. Also, our main criterion against which to measure success of a project is the 'opportunity cost of capital' (OCC), a minimum return in economic terms on capital in an alternative marginal use in the economy; the OCC can be used as a test rate of discount (TDR) to find the least-cost project or as a yardstick against which to measure the economic return (ER) to find if investing in the project at all is satisfactory intersectorally in the economy.

However, what we are comparing in this Appendix in economic terms is alternative projects for meeting the same load forecast (Chapter 4) at the same standard of security (Chapter 3), i.e. assuming that the benefits from the electricity are the same which ever project is chosen. We are concerned here with the least-cost solution and not the economic return.

For illustration we look at a common case for comparison today, i.e. a basically hydro development with a thermal development, for which the principle of comparison is simple but the actual arithmetic quite complex. This complexity requires a logical, systematic approach to avoid misconceptions and errors when doing the sums. To simplify matters even further we assume that the hydro plant is

not of the multipurpose type, i.e. it is built only for the production
of electricity. Such simplifications are needed for presentation
purposes only and not because of any shortcomings of the method
itself.

Other types of project can be compared by this same method, e.g.
gas-turbine generators versus steam turbogenerators, or nuclear
plant versus conventional plant.

8.5.2   Sequences of development

Although we are basically comparing a hydro plant versus a thermal
plant, both with their associated transmission, as we saw in Chapter
8 what we must really compare in practice is two separate sequences
of development:

(i)   The hydro alternative, typically consisting of a development
       of hydro plus thermal plant for back-up in case of drought,
       in our case with all the hydro plant introduced first.

(ii)   The thermal alternative consisting of a development of thermal
       plant only.

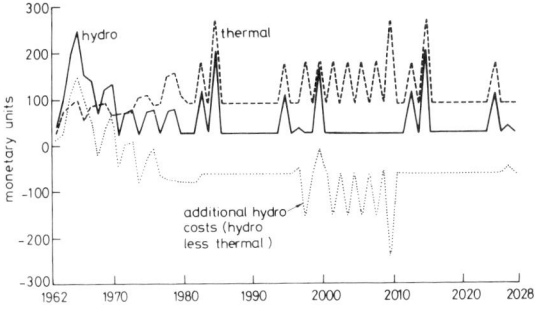

Fig. 8.3 Comparison of undiscounted cash flow of hydro and thermal
developments.

van der TAK, H.G.:'The economic choice between hydro-electric and
thermal power developments'. Chart 1. Published on behalf of the Johns
Hopkins University Press, Baltimore, USA, 1974. Reproduced by kind
permission of the Johns Hopkins University Press.

The cash flows associated with each alternative development in
constant price levels, i.e. allowing for only relative inflation,
are shown in Tables 8.4 and 8.5, respectively, and in Fig. 8.3; the
flows include both investment and operating (including fuel)
expenses. The hydro development has high investment expenditure,

Table 8.4 Cash flow of costs - Hydro Development

(monetary

Investment

Year	Dam	Hydro units	Trans- mission 400 kV	Trans- mission 230 kV	Trans- mission 115 kV	Thermal plant
Undiscounted						
1962	11750					
1963	23600			14000	2000	
1964	106200			25000	2000	
1965	147500			36000	2000	
1966	74900	51100				
1967	43300	57200		12900		
1968			38700			
1969		51100	51500			
1970		57200	52300			
1971						
1972		51100				
1973		57200				
1974		-				
1975		51100				
1976		57200				
1977		-				
1978		51100				
1979		57200				
1980		-				
--(1981)						
1983						90000
1985						180000
1995				77000	6000	
1997				12900		
2000			142500			
2013						90000
2015						180000
2025				77000	6000	
2027				12900		
2028						

Discounted

Present worth (PW) in 1962
of costs in period:

1962-1980		330580	265123	85443	74231	5201	0
Total 1981-2028 at 7.5%	0	0	9126	9032	615	59966
1962-2028		330580	265123	94569	83263	5816	59966
1962-1980		322035	243575	80046	72469	5108	0
Total 1981-2028 at 8.5%	0	0	6419	6473	442	47582
1962-2028		322035	243575	86465	78842	5350	47582

Source: van der Tak, H.G.: 'The economic choice between hydroelectric and
           World Bank by the Johns Hopkins University Press, Baltimore, USA,

*Operating and maintenance expenses of both hydro and thermal plant. Note:
merely financial or accounting such as depreciation, amortization and interest.
in which they are made. The hydro development results in a system using not only

units)

Operating expenses

Year	Pro- duction plant[*]	Fuel	Trans- mission	Frequency con- version	Total
Undiscounted					
1962	16440	9959			38149
1963	16440	11086		30000	97126
1964	16440	11763		30000	191403
1965	16440	13695		30000	247635
1966	14620	9296	830		150746
1967	14620	10704	830		139554
1968	14620	13624	959		67903
1969	16220	3532	959		123311
1970	16220	7540	959		134219
1971	16520	2052	2384		20956
1972	16520	2052	2384		72056
1973	16520	2052	2384		78156
1974	17520	2052	2384		21956
1975	17520	2052	2384		73056
1976	17520	2052	2384		79156
1977	18620	2052	2384		23056
1978	18620	2052	2384		74156
1979	18620	2052	2384		80256
1980	19320	2052	2384		23756
--(1981)	19320	2052	2384		23756
1983					+90000
1985					+180000
1995					+83000
1997					+12900
2000					+142500
2013					+90000
2015					+180000
2025					+83000
2027					+12900
2028	19320	2052	2384		23756

Discounted

Present worth (PW) in 1962
of costs in period:

		Pro- duction plant	Fuel	Trans- mission	Frequency con- version	Total
	1962-1980 ⎱	176168	78973	12112	78016	1105847
Total	1981-2028 ⎰ at 7.5%	63165	6709	7794	0	156407
	1962-2028 ⎰	239333	85682	19906	78016	1262254
	1962-1980 ⎱	165105	75999	10924	76621	1051882
Total	1981-2028 ⎰ at 8.5%	47279	5022	5834	0	119051
	1962-2028 ⎰	212384	81021	16758	76621	1170953

thermal power developments.' Table 1. Published on behalf of the
1974. Reproduced by permission of The Johns Hopkins University Press.

The cash flow of costs does not include any transactions that are
It includes only expenditures on goods and services in the year in
hydro but also some thermal plants.

Table 8.5 Cash flow of costs: Thermal development

Year	Dam	Hydro units	Investment Trans- mission 400 kV	Trans- mission 230 kV	Trans- mission 115 kV	(monetary Thermal plant
Undiscounted						
1962						
1963				14000	2000	
1964				25000	2000	
1965				38000	2000	
1966						29000
1967						59000
1968						59000
1969						59000
1970						30000
1971						29000
1972						30000
1973						29000
1974						59000
1975						59000
1976						30000
1977						29000
1978						88000
1979						89000
1980						30000
--(1981)						
1983						90000
1985						180000
1995				77000	6000	
1998						90000
2000						90000
2002						90000
2004						90000
2006						90000
2008						90000
2010						180000
2013						90000
2015						180000
2025				77000	6000	
2028						

Discounted

Present worth (PW) in 1962
of costs in period:

		Dam	Hydro units	Trans- mission 400 kV	Trans- mission 230 kV	Trans- mission 115 kV	Thermal plant
	1962-1980		0	0	0	65245	5201	326567
Total	1981-2028 at 7.5%	0	0	0	7888	615	94255
	1962-2028		0	0	0	73133	5816	420822
	1962-1980		0	0	0	63890	5108	296348
Total	1981-2028 at 8.5%	0	0	0	5667	442	70961
	1962-2028		0	0	0	69557	5550	367309

Note: The cash flow of costs does not include any transactions that are merely
only expenditures on goods and services in the year in which they are made.
Source: van der TAK, H.G.:'The economic choice between hydroelectric and thermal
            the Johns Hopkins University Press, Baltimore, USA, 1974. Reproduced by

units)	Pro- duction plant	Fuel	Operating expenses Trans- mission	Frequency con- version	Total	
Year	plant	Fuel	mission	version	Total	

Undiscounted

Year	Production plant	Fuel	Transmission	Frequency conversion	Total	
1962	16440	9959			26399	
1963	16440	11086		30000	73526	
1964	16440	11763		30000	85203	
1965	16440	13695		30000	100135	
1966	14620	9296	830		53746	
1967	14620	10704	830		85154	
1968	16180	12273	830		88283	
1969	17740	13382	830		90952	
1970	19300	16255	830		66385	
1971	20860	17895	830		68585	
1972	20860	19363	830		71053	
1973	22420	19836	830		72086	
1974	22420	21308	830		103558	
1975	23980	25434	830		109244	
1976	25540	29718	830		86088	
1977	27100	31542	830		88472	
1978	27100	34062	830		149992	
1979	28660	39904	830		158394	
1980	31780	42791	830		`105401	
--(1981)	33340	54631	830		88801	
1983						+90000
1985						+180000
1995						+83000
1998						+90000
2000						
2002						
2004						
2006						
2008						+90000
2010						+180000
2013						+90000
2015						+180000
2025						+83000
2028	33340	54631	830		88801	

Discounted

Present worth (PW) in 1962
of costs in period:

		Production plant	Fuel	Transmission	Frequency conversion	Total
	1962-1980 ⎫	206207	183062	5298	78016	870196
Total	1981-2028 ⎬ at 7.5%	109002	178611	2714	0	393085
	1962-2028 ⎭	315209	361673	8612	78016	1263281
	1962-1980 ⎫	191739	168133	5396	76621	807235
Total	1981-2028 ⎬ at 8.5%	81589	133691	2031	0	294381
	1962-2028 ⎭	273328	301824	7427	76621	1101616

financial or accounting such as depreciation, amortization and interest. It includes

power developments.' Table 2. Published on behalf of the World Bank by
permission of The Johns Hopkins University Press.

mainly for the dam, in the earliest years; the thermal development has high operating expenses mainly on fuel in the later years. Because of this difference in timing of major cash flows with each alternative, and to give a general time value to money, the figures in Tables 8.4 and 8.5 are discounted to a common year, say the year in which the sums are being done, using some discounted cash flow (DCF) technique.[15] This is illustrated by Table 8.6.

The cash streams are discounted at or around the OCC and the annual discounted values summed to a total present value of cost for each alternative, which can then be compared; the project with the lowest present value of cost (from any number of alternatives) is preferred.

Only actual cash flows are included in Tables 8.4 and 8.5. Interest and depreciation are taken into account by the DCF process itself; amortisation of any loans to supply the cash flows are a financial and not an economic matter, and are not included in the flows.

8.5.3 Period for present valuing

We first put down the cash flows for each alternative development over the initial period up to when present decisions do not influence the future; e.g. in our example to the point when the hydro development is fully utilised, i.e. 1980, when the last expenditure is made on hydro generators, see Table 8.4. However, the cash flows must be taken beyond this point until differences in the operating cost of each alternative, held constant at the level at the end of the initial period, become negligible in terms of their present value, i.e. 2028 in our present example. In order to know when to replace plant in the development sequence, economic lives are attributed to each physical item; e.g. transmission 30 years, gas turbine generators 15 years, diesel and steam turbogenerators 25 years, hydro turbo-generators 60 years, and dams 80 years. Investments in such project replacement are included in Tables 8.4 and 8.5.

8.5.4 Method of comparison

In general, as indicated above, the development which has the lowest present value of cost at the opportunity cost of capital is preferred and, if there are more than two alternative developments, the latter can be ranked in increasing present value of cost. Table 8.6 shows that at an OCC of 7.5% the hydro development is preferred, whilst at an OCC of 8.5% the thermal development is better. Somewhere around a discount rate of 8% the two developments are equal. One way to find out the equalising discount rate (EDR) between two alternatives is by inspection from Table 8.6. Another way is to subtract their cash flows, as shown by the dotted line in Fig. 8.3, and then to find the discount rate which makes the total present value of the new cash flow stream zero.

Whichever way we do it, we are trading off higher investment costs

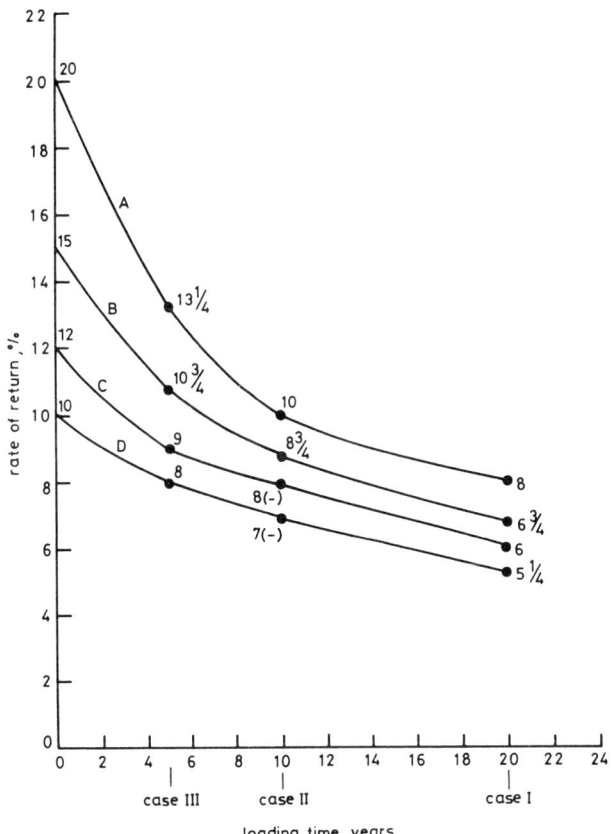

Fig. 8.4 Relationship between rate of return on investment in hydro
and length of loading time of dam

A : based on basic data
B : based on dam costs plus (nearly) 20%
C : based on dam costs plus (nearly) 40%
D : based on dam costs plus (nearly) 60%

VAN DER TAK, H.G.: 'The economic choice between
hydroelectric and thermal power developments'. Chart 3.
Published on behalf of the World Bank (Johns Hopkins
University Press, Baltimore, USA, 1974). Reproduced by
permission of the Johns Hopkins University Press

of one alternative development in the early years against higher
operating costs of the other alternative in the later years. As such
the EDR is a measure of the rate of return on the additional investment

Table 8.6   Summary of present worth of costs

| Cost component | Period 1962-2028 | | |
	Hydro develop- ment	Thermal develop- ment	Difference
			(present worth
Dam	331	–	331
Hydro units	265	–	265
Transmission 400 kV	95	–	95
Transmission 230 kV	83	73	10
Transmission 115 kV	6	6	–
Thermal plant	60	421	-361
Operating cost production plant	239	315	-76
Fuel	86	362	276
Operating cost transmission	20	9	11
Frequency conversion	78	78	–
Total	1262	1263	-1
			(present worth
Dam	322	–	322
Hydro units	244	–	244
Transmission 400 kV	86	–	86
Transmission 230 kV	79	70	9
Transmission 115 kV	6	6	–
Thermal plant	48	367	-319
Operating cost production plant	212	273	-61
Fuel	81	302	-221
Operating cost transmission	17	7	10
Frequency conversion	77	77	–
Total	1171	1102	69

Source: van der TAK, H.G.: 'The economic choice between hydroelectric
World Bank by The Johns Hopkins University Press, Baltimore,
Press

	Period 1962-1980		(thousands of monetary units) Period 1981-2028		
Hydro development	Thermal development	Difference	Hydro development	Thermal development	Difference
at 7.5%)					
331	–	331	–	–	–
265	–	265	–	–	–
85	–	85	10	–	10
74	65	9	9	8	1
5	5	–	1	1	–
–	327	-327	60	94	-34
176	206	-30	63	109	-46
79	183	-104	7	179	-172
12	6	6	8	3	5
78	78	–	–	–	–
1106	870	236	156	393	-237
at 8.5%)					
322	–	322	–	–	
244	–	244	–	–	–
80	–	80	6	–	6
72	64	8	7	6	1
5	5	–	1	1	–
–	296	-296	48	71	-23
165	192	-27	47	81	-34
76	168	-92	5	134	-129
11	5	6	6	2	4
77	77	–	–	–	–
1052	807	245	119	295	-176

and thermal power developments'.   Table 3.   Published on behalf of the
USA, 1974. Reproduced by permission of The Johns Hopkins University

Table 8.7 Present worth of costs: Comparison of higher and lower load
development, 1962-2028

	Hydro development		
	Lower load (1)	Higher load (2)	Differ- ence (2) - (1)
Cost component			(present worth
Dam	331	331	-
Hydro units	265	273	8
Transmission 400 kV	95	95	-
Transmission 230 kV	83	83	-
Transmission 115 kV	6	6	-
Thermal plant	60	60	-
Operating cost production plant	239	243	4
Fuel	86	92	6
Operating cost transmission	20	20	-
Frequency conversion	78	78	-
Total	1262	1280	18
			(present worth
Dam	322	322	-
Hydro units	244	252	8
Transmission 400 kV	86	86	-
Transmission 230 kV	79	79	-
Transmission 115 kV	6	6	-
Thermal plant	48	48	-
Operating cost production plant	212	215	3
Fuel	81	87	6
Operating cost transmission	17	17	-
Frequency conversion	77	77	-
Total	1170	1188	18

Source: van der TAK, H.G.: 'The economic choice between hydroelectric
World Bank by The Johns Hopkins University Press, Baltimore,
Press

	Thermal development		(thousands of monetary units) Difference (Hydro-Thermal)		
Lower load (1)	Higher load (2)	Differ- ence (2) - (1)	Lower load (1)	Higher load (2)	Differ- ence (2) - (1)
at 7.5%)					
-	-	-	331	331	-
-	-	-	265	273	8
-	-	-	95	95	-
73	73	-	10	10	-
6	6	-	-	-	-
421	441	20	-361	-381	-20
315	321	6	-76	-78	-2
362	406	44	-276	-314	-38
9	9	-	11	11	-
78	78	-	-	-	-
1263	1334	71	-1	54	-53
at 8.5%)					
-	-	-	322	322	-
-	-	-	244	252	8
-	-	-	86	86	-
70	70	-	9	9	-
6	6	-	-	-	-
367	388	21	-319	-340	-21
273	278	5	-61	-63	-2
302	340	38	-221	-253	-32
7	7	-	10	10	-
77	77	-	-	-	-
1102	1166	64	69	22	-46

and thermal power developments'. Table 6. Published on behalf of the
USA, 1974. Reproduced by permission of The Johns Hopkins University

in the hydro development.   Such an EDR should be equal to the OCC
to be acceptable.   The same sort of trade-off can be used for a nuclear
development versus a conventional thermal development or a conventional   thermal   development   versus   a   gas-turbine/diesel
development.

8.5.5 Sensitivity analysis of cost parameters

Tables 8.4 - 8.6 enable us to vary the main cost parameters and
do the sums again.   In our present example the following changes would
occur:

Change in parameter	Change in EDR (to)
None	8%
Dam cost down 20%	9%
Fuel price up 30%	9%
Hydro plant cost down 20%	8.5%
Thermal plant cost down 20%	7.5%

8.5.6 Sensitivity to shadow pricing

The same Tables enable us to carry out 'shadow pricing' on the main
parameters.   We shadow price the cash flows and do the sums again.

We need to know the foreign-exchange component to be able to do this
for the shadow pricing of foreign exchange.   In our example we assume:

Item	Foreign exchange component	Increase in cost resulting from 20% lower foreign exchange rate
	%	%
Dam	50	10
Hydro units	80	16
Thermal units	75	15
Transmission	80	16
Other equipment	0	0

In our example the EDR goes down to 7.5% using the above figures.
We can carry out a shadow pricing exercise on unskilled labour costs
in a similar manner; any other shadow pricing exercise will be
similar.

8.5.7 Sensitivity to load growth

When comparing any alternative developments it is essential to
check the sensitivity of the results against changes in load growth.

In order to do this we must rematch our generation development to
the new load forecast ending up with new versions of Tables 8.4 -
8.6 and Fig. 8.3. In our particular example the quicker the dam can
be fully used the more the hydro development is to be preferred, e.g.
if the date for the full utilisation of the dam is 1979 instead of
1981, as in our example, then the EDR increases from 8% to 8.25%;
(Table 8.7 refers). We can then carry out sensitivity tests as before
to get Fig. 8.4.

REFERENCES

[1] A good early reference is UK House of Commons Select Committee
     on Science and Technology, Second Report HC 381-XV11, 1967,
     particularly Appendix 44 and the Economic Annex, which are the
     important items for our purposes

[2] GITTINGER, J.P.: 'Compounding and discounting tables for
     project evaluation'. Published for the World Bank (Johns
     Hopkins University Press, Baltimore, USA, 1973). The interest
     rate is usually taken to be equal to the test rate of discount

[3] POSNER, M.: 'Fuel policy - a study in applied economics' (North
     Holland Press, 1973). Chap. 6

[4] BERRIE, T.W.: 'Appraising the economic worth of alternative
     projects', Elect. Rev., 29 Sept. 1967

[5] Using the interest rate for calculating the annual charges from
     the capital investment as the discount rate for present valuing
     of the individual annual load factors

[6] 'Central Electricity Generating Board'. Monopolies and Mergers
     Commission Report, HC 315, HMSO, London, 20 May 1981

[7] BATES, R., and FRASER, S.: 'Investment decisions in the
     nationalised industries' (Cambridge University Press, 1974)

[8] The author, for example, used the net effective cost method for
     some of his work with hydro-thermal systems in the World Bank,
     Washington DC, USA

[9] TURVEY, R.(Ed.): 'Public enterprise' (Penguin Modern Economics
     Series,1968) pp. 188-200

[10] Reference 6, Chapter 7

[11] TURVEY, R., and ANDERSON, D.: 'Electricity economics'.
     Published for the World Bank (Johns Hopkins University Press,
     Baltimore, USA, 1977)

[12] 'Power sector planning manual'.   UK Overseas Development Administration, 1979

[13] BRUCE, C.: 'Shadow pricing in project evaluation'. World Bank Staff Working Paper, 279, 1979

[14] This Appendix uses the method described in: VAN DER TAK, H.G.: 'The economic choice between hydroelectric and thermal power developments'. World Bank Staff Occasional Paper Number One (Johns Hopkins University Press, Baltimore, USA, 1974). The Tables and Figures have been taken directly from this Occasional Paper by permission of the Johns Hopkins University Press.
    Although the actual amounts of cash shown are out of date, this does not distort in any way the illustration of the method. Although the reader must remember that the balance of fuel to capital costs in 1982 would be different from that shown, once again, this does not distort the illustration of the method.

[15] GITTINGER, J.P.: 'Compounding and discounting tables for project evaluation'. Distributed for the World Bank (Johns Hopkins University Press, Baltimore, USA)

Chapter 9

Pricing and load management

9.1 MARGINAL-COST PRICING

9.1.1 Introduction

In the past electricity prices[1] were average-cost based. The tariff level was determined from financial targets set by government, regulatory body, or the utility itself, using historic costs from the utility's accounting books. The tariff structure ensured each consumer paid his own 'share' of the average cost of supply, e.g. as determined by how much of the system capacity and fuel each consumer class used on average.

Average-cost pricing gives few 'signals' about future financial resources needed for increasing electricity requirement or new consumers. Marginal-cost pricing overcomes this important defect; best known for its early application in France and Britain, it is now the declared policy of most power utilities in the world. Lending agencies, most government departments and regulating bodies insist that marginal-cost pricing is the proper basis for tariffs.

In marginal-cost pricing[2] past or 'sunk' costs are not significant; only uncommitted costs incurred because of additional demands provide the correct 'signal' to consumers. These are the investment and running costs incurred when consumers demand more. Marginal-cost pricing was the first step uniting producers and consumers in a joint optimisation of their objective functions. Marginal costs are usually related to extra increments of 1 kW or 1 kWh of electricity demand.

9.1.2 Tariff components

Marginal-cost tariffs have the same components as average-cost tariffs but with different numerical values:

Component 1: Capacity (kW) charge(s), related to the system peak demand

Component 2: Management, operating and maintenance charges

<u>Component 3:</u> Fuel (kWh) charges, related to system fuel costs

<u>Component 4:</u> Connection and metering charges

Component 1 plus the fixed part of component 2 are 'capacity related', i.e. directly concerned with an incremental kW. Component 3 plus the variable part of component 2 are 'energy related', i.e. directly concerned with an incremental kWh. Component 4 is 'consumer related', i.e. directly concerned with the consumer. Marginal-cost tariffs are calculated according to the same consumer classes as average–cost tariffs, basically residential, commercial and industrial.

9.1.3 Capacity-related charges

To determine the marginal 'capacity-related' charges we calculate the capital and fixed running costs of installing new plant (possibly delaying scrapping old plant) to meet an increase in total peak demand of 1 kW. We take the new generating plant as either (a) the most recent 'mix' by fuel and size or (b) the predominant plant type and size being currently installed. We include the capital cost of the attributable network; this varies with the consumer, e.g. with connection voltage. Allowance is made for capacity (kW) losses; 1 kW delivered requires more than 1 kW output from the generator. It is often necessary to plot a trend from past incremental expenditure on the network to obtain these incremental capital costs; the attribution of these costs between consumer classes can also be done by trending.

Only consumers increasing peak demand normally lead to capital costs; thus no capital charges are levied on 'off-peak' loads under present marginal-cost pricing. This was not necessarily the case under average–cost pricing. Capital charges are reduced by any fuel-cost savings made by new plant (see Chapter 5).

9.1.4 Energy-related charges

Energy-related charges include fuel and the variable component of management, operating and maintenance costs of the system to meet an incremental kWh at a particular hour and location. Whole-system costs must be taken and not just the energy-related costs of new plant because an interconnected system operates as one entity to meet any extra kWh.

9.1.5 Consumer-related charges

Consumer-related charges are for connections, metering and billing costs. These are often found by trending in the same way as network costs, as described above.

9.1.6 Practical tariff-making

Marginal fuel costs at times near system peak demand are many times those at off-peak, but in practice it is difficult to reflect this in tariffs; it could mean increasing existing peak tariffs by a large amount.

Also, charging purely marginal costs gives no guarantee that the utility's financial targets will be met in the short term, (although they may be in the long term), especially if capacity and energy-related charges are derived from a development plan based on 'shadow-pricing' (see Chapter 2). To adjust marginal-cost pricing to meet the financial target and take account of other practical difficulties we must:

(i)   progress smoothly from existing to future tariffs
(ii)   adjust for metering costs when these are high
(iii) take into account that only large consumers understand these
       tariffs unless a direct display of prices being charged is
       available.

Load management and spot pricing are a direct consequence of using marginal-cost tariffs. If consumers will reduce their load at crucial times for the utility, this saves incremental costs which can be passed on to such consumers. Autoproducers can also play a truer part in the power system under spot pricing than they can under existing tariffs (see later).

Marginal-cost tariffs can be costly to administer and need to have a consumer response. They are most suited to direct application in developed countries for large consumers. In developing countries and for small consumers, the capacity charge is usually incorporated into the energy charges to make the tariff easier to understand and implement.

In poor countries there may be a low 'lifeline' price for special consumers using very little electricity but who would be unable to sustain life without that small quantity.

Increases in fossil fuel costs are usually allowed for by a fuel-cost adjustment clause to adjust automatically the tariff upwards, at least for bulk supplies.

9.1.7 Commercial contribution to tariff making

Several major disciplines contribute to practical tariff making. The 'commercial' approach uses the experience of the 'commercial' engineer whose knowledge of the electricity market is important; how consumers will react to change; which consumer classes are the most important to cultivate at any particular time; and what the market will bear in tariff increases.

9.1.8   Accounting approach to tariff making

The accountant is needed to ensure the short- and medium-term financial viability of the utility and to make full use of the utility's complex accounting documents for tariff making.

9.1.9   Economist approach to tariff making

The economist's view is shared by the planner but may seem foreign to the commercial and accounting approaches. For example, only economists can easily see that we should set tariffs according to economic resource costs and not money costs; because, in the long term, what is good for the economy must be good for the utility and ultimately for the consumer.

The short and the medium term are most important to the accountant and commercial engineer, while the long term is most important for the economist. The economist changes the 'money' costs taken from the accounting books into 'economic' costs by 'shadow-pricing', largely to reflect differences between the utility and the national economic viewpoints. For example, taxes are part of money costs but are only internal transfers to the economy; thus economic costs exclude them. Also economic costs must be weighted to reflect scarcities or excesses in resources, e.g. foreign exchange, local capital, fuel, skilled labour etc. Economic costs are specified in constant price terms taking account only of relative inflation between different types of costs, e.g. between capital and running costs or different kinds of running costs. Any tariff structures recommended by the economist will be based on the above concepts and on the long-run marginal costs of a development programme under the investment, pricing and reliability rules of Chapter 2.

9.1.10   Putting the contributions together: the overall strategy

Putting together the contributions from the three disciplines means:

(i)   The economist working out an ideal long-run marginal cost tariff, the commercial and planning engineers providing cost inputs, i.e. for the fuel and other running costs of the utility, and the investment costs for the development programme.

(ii)   The accountant examining the utility's financial projections for the short and medium term to determine the required 'pitch' of the tariff level, with the aid of the commercial engineer to allocate investment and running costs to the various consumer classes, thus determining short- and medium- term tariffs which will be equitable, and cover the utility's marginal costs and financial obligations.

(iii) The commercial engineer examining the existing tariff to determine obvious cases where consumer classes are presently over- or undercharged, also to decide whether the overall level of the existing tariff is commercially satisfactory.

(iv) All disciplines producing a tariff based as closely as possible on the economist's marginal-cost tariff, and proceeding from the existing tariff to the future tariff as smoothly as possible.

(v) The process might well be iterative.

9.2 LOAD MANAGEMENT

9.2.1 Introduction

Electricity is a 'premium' fuel, i.e. a fuel in a most-refined grade compared for example with oil or coal. Therefore conserving[3] and optimising usage of energy applies especially to electricity. In recent years, because fuels are costly and for environmental reasons, consumers have become accustomed to government 'interference' in the use of electricity.

Because electricity cannot be stored, except at prodigious cost, and in order to produce and use it more efficiently it is important to improve the overall load factor of the system and so optimise generating-station production. Improved loading on the system results in more efficient use of existing plant, and reduces the need for new plant.

Load management results in lower costs to the utility which can be passed on via tariffs to consumers, and especially those who participate in load management schemes.

9.2.2 Control of peak- and off-peak usage

Load management as applied to peak- and off-peak usage acts as a useful 'bench mark' to measure the efficient use of power systems and fuel purchases. Before marginal-cost pricing was accepted there were several ways of discouraging the use of electricity at times of peak, e.g.:

(i) Special terms for assured off-peak supplies; either a load could not be physically connected at peak time or the consumer paid a penalty if it were connected then.

(ii) Special, often complex, commercial agreements between utilities and large commercial or industrial consumers entitling them to lower charges outside the peak period.

(iii) A tariff encouraging the continuous use of the system

Table 9.1   Average retail rates in residential tariffs in England and Wales
(Constant price levels: 1976)

Type of tariff	15 hour duration (12 hours night+ 3 hours mid-day) (p/kWh)	11 hour duration (8 hours night+ 3 hours mid-day) (p/kWh)	8 hour duration (Night only) (p/kWh)	Running rate in standard residential tariff (p/kWh)
Financial year				
1956/57	1.25 (87)	–	–	1.43 (100)
1966/67	0.93 (55)	0.83 (49)	0.76 (45)	1.69 (100)
1976/77	1.41 (62)	1.27 (56)	1.14 (50)	2.28 (100)
1986/87*	2.72 (77)	2.34 (67)	2.21 (63)	3.50 (100)

* Estimated by the author of this book from published data
() Shows the off-peak rates expressed as a proportion of standard
residential tariff running rate in last column

Source: Platts, J.: 'Britain's experience of electricity
marketing directed towards load management,' Paper
to energy conservation and the utilisation of off-peak
power seminar, Washington DC, Sept. 1978.

throughout the year, once the capacity-related charges had been recovered.

(iv)   Promoting a more flattened demand against time characteristic, by selling appropriate apparatus, e.g. off-peak heating or cooling equipment instead of peak-load radiant fires, air conditioners and refrigerators.

(v)   Concessional terms for off-peak supplies. This obverse of peak-load pricing attempted to 'fill in' any troughs in the power-system demand against time profile. Because of delayed response several systems have over-filled a trough, causing an awkward new peak. This must be avoided, usually by not trying to fill in the troughs completely, i.e. by deliberately leaving a margin of the troughs unfilled to allow for uncertainty in off-peak load pick-up.

9.2.3   Load management by pricing[4]

Over the last 25 years special peak and off-peak tariffs have been greatly developed. The earliest examples were in France and the UK. Table 9.1 shows their broad development. Changes in the general relative tariff level are due mainly to fuel cost changes.

Table 9.1 shows how inducements have been made to encourage the use of off-peak electricity and discourage its use at peak times. In England and Wales Fig. 9.1 indicates that the main result of this has been to fill up the winter night 'valley'. By 1976-77 the demand at 0200 hours had doubled compared to 1956-57. By 1986-87 the demand might be 2.5 times as much as in 1956-57.

Table 9.2 gives some indications of the effect on annual system load factor.

Table 9.2   Annual system load factor in England and Wales

Year	Annual system load factor	Proportion of total energy supplied at night
	%	%
1956/57	48	19
1966/67	52	22
1976/77	57	25
1986/87*	59	28

* Estimated by the author of this book from published data

Source: PLATTS, J.: 'Britain's experience of electricity marketing directed towards load management'. Paper presented to Energy Conservation and the Utilisation of Off-Peak Power Seminar, Washington DC, Sept. 1978. Reproduced by permission of the author.

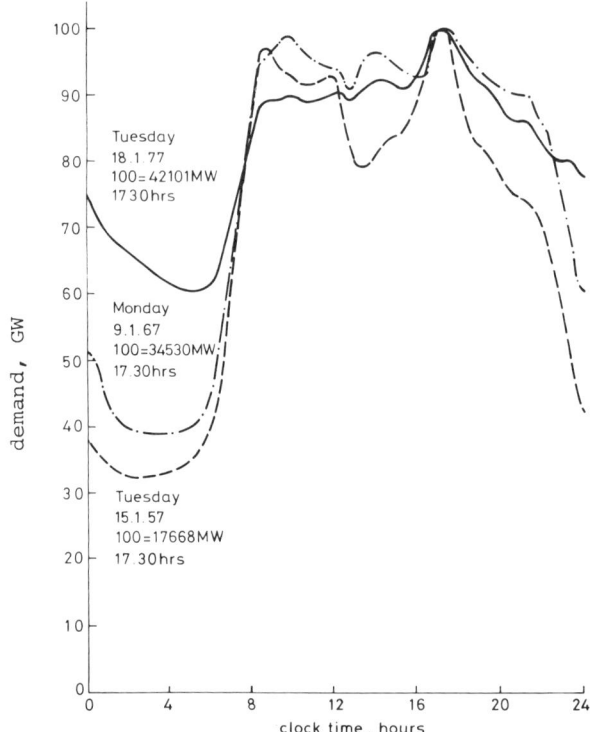

Fig. 9.1   Day of maximum demand on CEGB 1956/57, 1966/67, 1976/77,
adjusted to the same level of system peak demand at 1730
hours

PLATTS, J. : 'Britain's experience of electricity marketing
directed towards load management'. Paper to Energy
Conservation and the Utilisation of Off-Peak Power
Seminar, Washington DC, Sept. 1978. Reproduced by
permission of the author

Figs. 9.1, 9.2 and 9.3 illustrate how the system demand against
time curve has levelled out over the year in England and Wales,
indicating continuous success in load management.

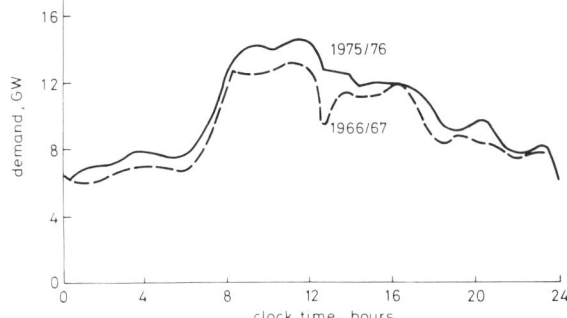

Fig. 9.2 Midwinter weekday load curves. Total industrial load (at
0°C) met by Boards in England and Wales, adjusted to the
same level of peak demand at 1600 hours

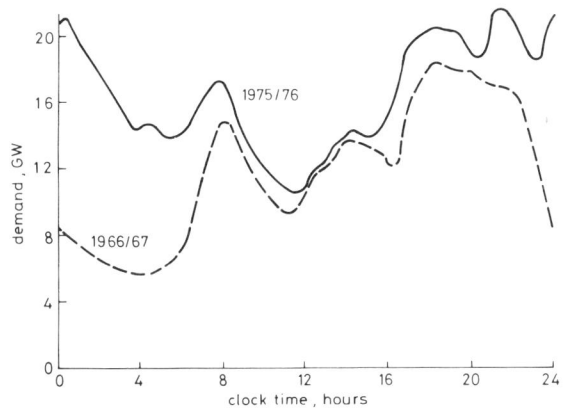

Fig. 9.3 Midwinter weekday load curves. Total domestic load (at
0°C) met by Boards in England and Wales adjusted to the same
level of peak demand at 1200 hours

PLATTS, J.: Figs 9.2 and 9.3 from 'Britain's experience
of electricity marketing directed towards load management'. Paper to Energy Conservation and the Utilisation
of Off-Peak Power Seminar, Washington DC, Sept. 1978.
Reproduced by permission of the author

Table 9.3   Rates for additional demand of 2MW at 50% load factor in specimen industrial high voltage tariffs in England and Wales

Year	Demand charge £ per kW per month		Running rate (including fuel-cost adjustment) p per kWh		Average price for 50% load factor supply p per kWh
	Winter	Summer	Day	Night	
1956/57	1.13		1.27		1.55
1966/67	2.06		1.01	0.87	1.48
1976/77	1.71	0.65	1.57	0.92	1.72
1986/87*	2.34	0.92	2.32	1.34	2.42

* Estimated by the author of this book from published data

Source: 'A review of off-peak tariff incentives and their effect in England and Wales'.   Prepared by J.A. Burchnall for UK Government, Symposium on Load-Curve Coverage in Future Electric Power Generating Systems, Rome, Italy, Oct. 1977; organised by ECE.   Reproduced by permission of the author

9.2.4   Peak-load management by pricing

Only large loads warrant the expense of complicated metering and complex explanatory leaflets which go with special peak-load tariffs, but in these cases load management by such means is well worth while. Just as off-peak load control by pricing filled in the troughs in the demand against time curve, peak-load control by pricing is designed to flatten the peaks. To keep the illustrations consistent, the example of peak load management by pricing has again been taken from England and Wales, using a broad 30 year spectrum, with the 1986-87 figures estimated from published data.   Table 9.3 gives an illustration of peak-load tariffs used for this purpose.   Another illustration is that of reducing the capacity charge for large industrial consumers who agree to take a reduced load at peak if requested by the power utility.

9.2.5   Load control by direct means[5]

Consumers can thus be persuaded by pricing to instigate load management on their own behalf. However, load management can also be affected by direct physical control. This has existed for just as long as peak/off-peak pricing incentives, and probably for longer in the shape of the 'current limiter'. This is a relatively simple device to disconnect automatically all or part of a consumer's network when the total demand reaches the specified cut-off point as determined by the consumer and utility. The higher the cut-off point, the higher the price which the consumer must pay for supply. Such load limiters are still used today, especially in developing countries, but are regarded as crude and somewhat erratic in operation; there are now more sophisticated devices for fulfilling the same function.

Such sophisticated controls fall into three groups:

(i)   time control devices
(ii)   manual control devices
(iii) centralised control devices.

Time control devices switch loads off and often on again, usually according to a strict timing sequence based upon a previous estimate of the likely time of peak load. By its nature time control is most suitable for loads with a consistent cyclic pattern of consumption, e.g. day-to-day.

Manual control devices do the same job as time control devices, but do it manually and are based directly on human judgment. Usually someone determines the cut-off time by observing a demand (kW) meter. Using prescribed rules or judgment alone, the observer can then physically disconnect either some or all of a consumer's load in order to prevent the system maximum demand from exceeding a predetermined amount.   The observer can also usually restore the load(s).   Skill is needed in using this method if under-shooting or over-shooting is to be avoided. However, at times of chronic shortages in some countries it can be used with a fair measure of success because the operator soon becomes experienced. A sophistication is to use two operators, one at the consumer's premises, the other at the utility's premises. These operators are in contact one with another, e.g. by radio, telex, a light signal, or by telephone. Two operators are less likely to under- or over-shoot and can load manage with a good deal of precision.

9.2.6 Centralised control

Systems with many different loads of varying types can carry out load management most efficiently by using centralised systems, nowadays usually electronic. Normally maximum demand (kW) and electrical energy (kWh) are monitored simultaneously. Such a degree of sophisticaion makes possible complicated contractual agreements with consumers as to the timing of cut-offs and restorations of supply.

All three controls are intended to switch loads both off and on. However, some loads cannot be switched off without considerable financial loss to the consumer, e.g. glass and steel works; others are easily switched off and on without fear of damage since operations are shifted from peak to off-peak. The following groups are suitable for the switching-off/switching-on process:

Arc furnaces	Incinerators
Battery chargers	Lighting
Compressors	Motors (small)
Fans	Pumps
Furnaces	Space heaters
Grinders	Water heaters

Large motors should not normally be considered for switching off and on because of the high starting currents and mechanical stresses involved.

For centralised load management, data must be collected into a centralised control system to monitor energy consumption (kWh) and maximum demand (kW). The desired demand level is usually termed the 'target' or 'control' level. Again, there will inevitably be over-shooting and under-shooting of the 'target' at any instant; also under-shooting will normally be over-corrected by over-shooting and vice versa. In this respect, using a computer to work out past trends of growth in demand for comparison with the present can prove useful for systems with either a large number of load centres or a number of large load centres.

Today most utilities are capable of load management on a large scale using signals sent out from a central point, e.g. a power station or a substation. There are three types of channel for sending these signals:

(i)   telecommunication circuits
(ii)   radio transmitters and receivers
(iii) electricity supply conductors themselves.

Unless a utility is already well equipped with these, the use of telecommunication circuits is usually prohibitively expensive to install and maintain just for load management purposes. A one-way radio control system requires a transmitter at the centralised control point, a receiver at the load point, and an actuating device to switch the load off/on. A two-way device requires a transmitter and a receiver at each end. Coded signals can then be used to discriminate between types of load. One disadvantage of this method is the limited number of radio transmission frequencies usually available. However, radio control of residential water heaters, air conditioners and similar loads has been in use for some time.

When using the electricity supply conductors themselves, signals are injected into the network via coupling circuits usually at a substation. A programmed unit is used to send out the signals at the proper time; these are decoded at the load point, decoupled from the network, and used to actuate the load management device. A good summary of centralised control load management systems is given in Reference [2].

9.3   INTERACTIVE LOAD CONTROL

9.3.1   Introduction

Marginal-cost tariffs based upon a prescribed load forecast met by a least-cost development programme have had mixed success in achieving either optimal load forecasts or optimal development programmes. There are too many variables to be estimated accurately, e.g. the weather, consumers' demands, fuel costs and plant availability.

Recent events in the microprocessor and communications sectors make it possible to set prices virtually at the time the electricity is consumed rather than in a predetermined way. These new 'spot' prices are based on the actual demand and plant conditions at the time the price is set.[6] Spot prices can be updated as often as every five minutes.

On a warm winter day a spot price would be lower than on a cold winter day, not possible under normal prescribed pricing and load management rules. The consumer and producer share any good or bad fortune. If the producer has more plant than expected the spot price will be lower and the reverse will be true for having less plant than expected. Spilling of water over dams and blackouts should both be avoided by spot pricing. Autoproducers and cogenerators will always receive a buy-back price equal to the spot price, i.e. generally much higher than at present (these buy-back prices will be the same for each type because, although autogenerators may have no heat output 'per se', both have electricity output).

Large consumers who make up a substantial part of demand covered by load management schemes are obvious choices for spot pricing because they will be more likely to pay for new measuring apparatus and to understand the 'message' in a spot price. Experiments are also proceeding using spot pricing with residential consumers, e.g. in the UK.[7]

9.3.2 Spot pricing

Tariffs with some form of spot pricing have been in use for some time.[8] Spot pricing has been offered as a surcharge during periods of likely plant shortage; in Britain this is a 'peak-load warnings' surcharge and applies to many hundreds of consumers.[9] Illinois Power and Light Company offers spot pricing as an alternative to disconnection when plant is short.[10] Many utilities have supported spot pricing for electricity demand/supply control.[11]

Spot pricing established for the first time a true 'market place' enabling consumers and producers to adjust demand and price simultaneously. It is workable whether a utility is publicly or privately owned, centrally or locally owned/controlled. Auto-generators and cogenerators are equal partners with mains generation and we abolish any concepts of 'firm' or 'unfirm' supplies, e.g. from renewables. Electricity from any source has a value/price at every instant it is available. Spot pricing copes fairly with electricity being a monopoly supply unless it is inefficient.

9.3.3 The simple case

Here we only illustrate the principles behind spot pricing and indicate its likely success; both the philosophy behind it and its application are in their infancy. We start by making some simplifying assumptions:

(i)   All main generators are owned and operated by the same centrally controlled utility.

(ii)   All capital costs and the cost of transmission and distribution losses are ignored.

(iii)   Demand is never greater at any instant than total generation capacity.

(iv)   The utility chooses the level of capital investment only once and at the start of the exercise, after which the number, type and size of the generating units remain constant.[12]

(v)   Time-series characteristics have second-order effects on demand and operating costs, an assumption already used extensively in planning.

(vi)   Generation start-up, shut-down and increase/decrease rates are costless and not constrained from one period to the next.

(vii) Only present consumption determines the output function of a consumer, not past or previous consumptions.

(viii) Processes like storage are not considered except in some very obvious ways[13], e.g. hot-water storage, over-night heat-storage units.

(ix) Spot demands in each time period are independent.

(x) System running costs in each time period are independent.

The appropriate criterion to use for optimum decision-making is global welfare optimisation, i.e. maximising the difference between the total value 'per se' of consumption to the economy and consumers and the cost of the generation, subject to this being available. The spot price derived from this welfare optimisation is a 'buy' price for a net consumer of electricity at any time, and a 'buy-back' price for a net producer, i.e. production and consumption are treated symmetrically.

9.3.4 Simple graphical solution to the problem

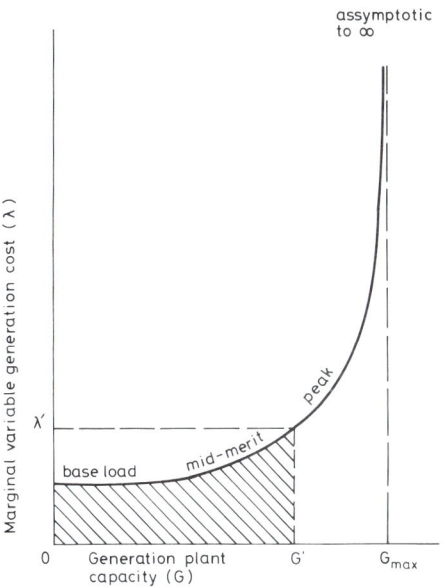

Fig. 9.4 Marginal variable generation costs for fixed plant capacity and plant availability

> BERRIE, T.W.: 'Interactive load control: Peak load pricing and investment', <u>Elect. Rev.</u>, 9 Oct. 1981. Reproduced by permission of <u>the Editor</u>

We give the solution to this simple case by graphical means. Fig. 9.4 shows the system marginal variable generation cost λ against generating plant capacity G for a fixed total capacity of plant and a fixed plant mix. The curve is always level or upward-sloping because generating units are always loaded in order of increasing variable marginal cost. The system marginal variable cost becomes infinite at the point G_{max}, the vertical line in Fig. 9.4.

The cumulative generation cost $T(G)$ is the area under the curve up to the value of G representing the amount of the generating plant committed at that instant. Thus $T(G_{max})$ is the total shaded areas under the curve.

This is the picture from the generation or supply side. Fig. 9.5 shows an analogous picture from the consumer or demand side, the well-known economist's demand curve. The consumer will set his demand at a level where his marginal value of productive output per unit equals the price, that is, at demand level D' for price level p' on Fig. 9.5. Then according to the normal economist's demand curve, the shaded area in Fig. 9.5 is the value of the consumer's production.

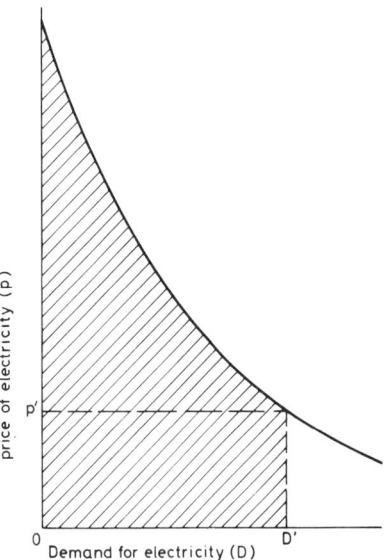

Fig. 9.5 Demand and consumer production worth for prescribed demand

BERRIE, T.W.: 'Interactive load control: Peak load pricing and investment', <u>Elect. Rev.</u>, 9 Oct. 1981. Reproduced by permission of the <u>Editor</u>

Optimal spot prices are obtained from superimposing Figs. 9.4 and 9.5 to give Fig. 9.6. If we set prices at p' we know from Fig. 9.5 that demand will be D', and the shaded area in Fig. 9.6 is the net value of the economist's short-run welfare function represented by the area:

(i)   below the demand curve D, on the demand side; but

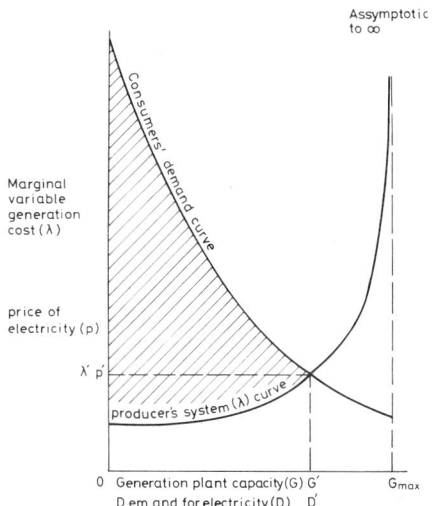

Fig. 9.6   Short-run welfare value

BERRIE, T.W.: 'Interactive load control: Peak load pricing
and investment', <u>Elect. Rev.</u>, 9 Oct. 1981. Reproduced by
permission of the Editor

(ii)   above the marginal-cost curve λ, on the supply side.

As such the shaded area shows:

The value of the consumer's output production per unit of
electricity at the margin;

<u>less</u>

The value of the producer's cost per unit of electricity at the
margin.

Analytical methods used in Reference 6 show that the case we present
here is not materially altered in the more normal case of multi-period
investment; i.e. assumption (iv) above is not binding.
    We can now go on to find the optimal 'spot price'. Fig. 9.6 shows
spot pricing under normal conditions when there is adequate capacity

to meet demand. The spot price p' is exactly equal to the short-run marginal cost of production λ', i.e. at point p' for demand D' at any instant in time. Fig. 9.7 shows the solution for spot pricing when all available capacity including those of autoproducers is fully used. This time p' must be set, in accordance with our previous assumptions, so that demand never increases beyond G_{max}. Thus we have a 'premium' component in the spot price, extra to the short-run marginal cost of production to ensure that demand is cut back to equal supply. In the absence of this premium, load shedding would take place. Unexpected lower availability of plant has the same effect as higher than expected demands. The higher the demand or the lower the plant availability than expected, the larger the shortage 'premium' on the spot price.

9.3.5 Spot pricing and the investment rule

We need to decide how our investment rule of Chapter 2 should be altered to accommodate spot pricing. To do this we relax our assumption that spot prices are derived from a once-and-for-all quantity and mix of generating plant, with the realised values for demand and plant availability as random variables. We now select the quantity and mix of generating plant to maximise total economic welfare as before, but this time take capital costs into account.

If we add one plant item P on to the system, its effect is shown in Fig. 9.7. The value of the added plant P depends upon the difference between the expected demand and actual demand, and the actual availability of the additional plant. When not available the additional plant contributes nothing to the spot price; when available the spot prices will be affected as shown in Fig. 9.7. The shaded area is a graphical illustration of how the additional generating plant contributes to the short-run welfare and spot pricing. The 'premium' to keep demand within capacity limits is reduced by the introduction of the additional plant; no premium at all may be needed as shown in Fig. 9.7.

In all cases there may well be a reduction in the total system fuel costs due to the introduction of the additional generating plant P because that plant is more efficient than existing plant. Thus building additional plant P has two components from which we can deduce our investment rule:

Component one: The 'premium' needed to keep demand within the bounds of total generation capacity available is reduced whenever plant P is available. This is shaded area B in Fig. 9.7.

Component two: The fuel cost of generation for the optimum spot price demand D in Fig. 9.7 will be reduced whenever plant P is operating and is not the last generator in the merit-order. This is area A in Fig. 9.7.

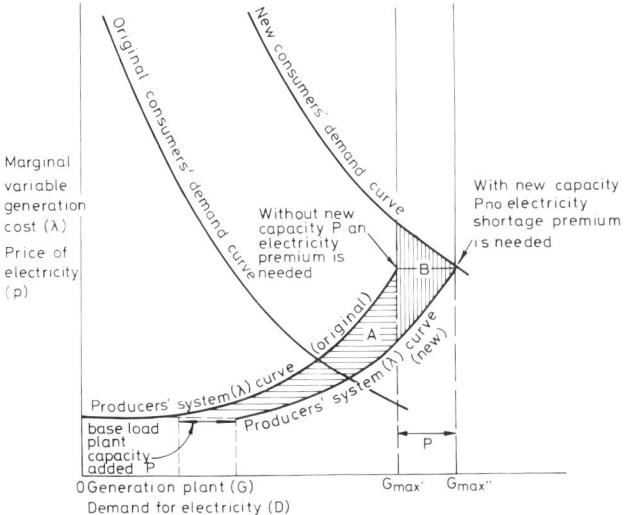

Fig. 9.7 Spot pricing under normal conditions

BERRIE, T.W.: 'Interactive load control: Peak load pricing and investment', <u>Elect. Rev.</u>, 9 Oct. 1981. Reproduced by permission of the Editor

The total economic value of adding plant P is the summated value of these two components over the life of plant P. To examine what our investment rule should be under spot pricing we must determine the total contribution to short-run welfare of adding an increment of plant P, as measured by the shaded areas in Fig. 9.7 over all possible values of demand and plant availability. We then compare this total contribution with the increased capital cost of plant P. If the total contribution to short-run welfare is greater than the increased capital cost, then we should increase the capacity of plant P until there is no difference.

Spot pricing thus brings our three rules of Chapter 2 closer together: in one sense it integrates them; in another it puts the fundamental emphasis upon the pricing rule. Sectors other than electricity will be in immediate rapport with this view, because price is at the heart of their planning and operation. We can therefore look forward to a body of literature on pricing and investment in electricity supply, using spot pricing, nearer to the traditional form of other sectors, i.e. starting with and ending on price.

9.4 SUMMARY

The traditional average cost pricing in the electricity supply industry does not give an accurate signal to the consumer concerning the economic resources needed to supply him with an extra kWh at a particular time-of-day/week/month/year.

Marginal cost pricing comes much closer to giving the right signal, although a theoretical marginal-cost tariff structure might well have to be modified for practical reasons concerning metering apparatus, consumer comprehension and the financial viability of the power utility.

Load management schemes enable both the consumer and the power utility to plan ahead more accurately and make for lower-cost electricity production. Interactive load control between electricity consumers and producers, operating very near to real time, enables prescribed tariffs to be done away with, their place being taken by 'spot pricing', in a true market place atmosphere linked with consumers' load management schemes, spot pricing should make for lower electricity production costs and allow, through realistic buy-back prices, for autoproducers and cogenerators to play their full part in the interconnected power system.

9.5 APPENDIX A: CALCULATING IDEAL LONG-RUN MARGINAL COSTS

9.5.1 Introduction

We define ideal long-run marginal costs (LRMC) as the incremental cost of the adjustments to the power-system development programme, and in power system operation attributable to a sustained incremental demand increase occurring at some point in time which is not necessarily at time of system peak demand. What we mean here by the word 'incremental' is a very small, discrete change interpreted in the strictly mathematical sense, with reference to the infinitesimal change of the differential calculus, i.e. approaching zero.   In practice we assume it to be a 1 kW change. The structuring of LRMC is based chiefly on such things as time-of-day, season, voltage level and geographic area.   The degree of sophistication of LRMC depends on data constraints and the usefulness of the end result. Ideally the LRMC of each individual consumer at each moment of time can be determined.   A very full account of the calculation and practical modification of ideal long-run marginal cost tariffs is given in several publications; see Reference 1.

9.5.2 Capacity-related costs

Ideal marginal capacity-related costs are the investment and fixed operating costs of generation, transmission and distribution facilities for supplying one additional kilowatt when this occurs at time of system peak. Ideal marginal energy related costs are the fuel and variable operating costs of providing the energy associated with the additional 1 kW of demand when this occurs at any hour; they are usually measured per kWh. Ideal marginal consumer related costs are the costs directly attributable to one consumer, including costs of connection, metering and billing.

In Chapter 5 we saw that an examination of the power-system load-duration curves enables us to determine time periods likely to be near to the peak at a particular time of day, or at a particular season of the year. Let us first make the idealised assumption that we have only two time periods, viz. 'peak' and 'off-peak'. We now build on this simple system, ignoring the difficulties which this assumption makes; for example, if the duration of the peak period is too narrowly defined, pricing is likely to cause a shift in the peak to the off-peak.

9.5.3 LRMC capacity-related costs

Fig. 9.8 shows the annual load duration curve (LDC) ABE for the system in year 0, and for the two simple peak and off-peak periods. The LDC increases in magnitude with system growth. The peak demand forecast is given by curve D in Fig. 9.9 starting from the initial value MW_0.

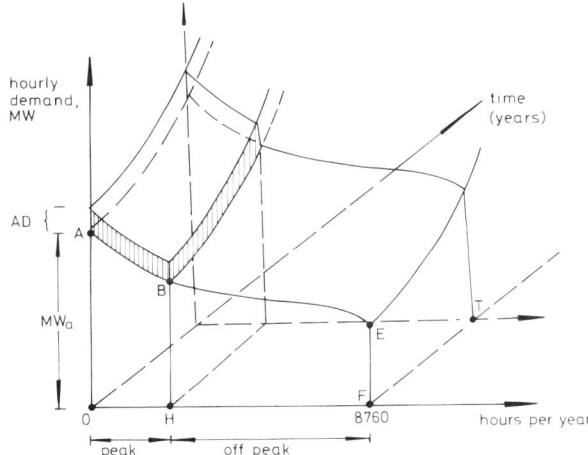

Fig. 9.8   Typical annual load-duration curve (LDC)

'Electric Power Pricing Policy'. World Bank Staff Working Paper 340, 1980. Reproduced by permission of the World Bank

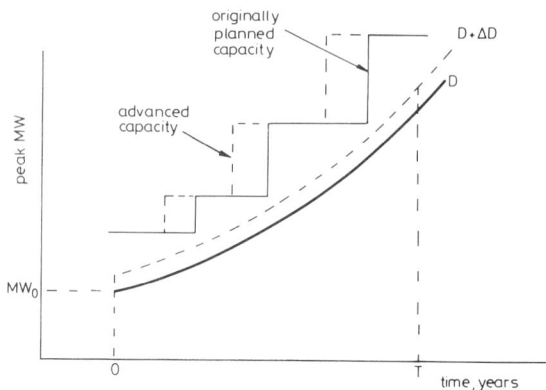

Fig. 9.9   Forecast of peak power demand

'Electric Power Pricing Policy'. World Bank Staff Working Paper 340, 1980. Reproduced by permission of the World Bank

The LRMC capacity-related cost is the change in system capacity costs ΔC associated with a sustained increment in the long-run peak demand shown by the shaded areas of Fig. 9.8, and the broken line $D + \Delta D$ in Fig. 9.9. The LRMC of generation is given by ($\Delta C/\Delta D$), provided that ΔD is marginal <u>both</u> in time, and in terms of kW. Ideally ΔD can be either incrementally positive or incrementally negative.

Fig. 9.9 adopts the growing philosophy amongst power-system planners today that an extra kW of peak incremental load would normally be met by advancing future base-load generating plant. If this is not operationally possible it would be met by inserting peaking units such as gas turbines or peaking hydro plant. In our planning we then calculate ΔC by simulating the system expansion and system operation, with and without the extra kW demand increment ΔD. Let us suppose that in our base case gas turbines are used for peaking rather than base load plant. The required LRMC of generating capacity is then the cost of advancing 1 kW of gas turbines in terms of the cost per kW installed annuitised over the economic lifetime. This figure must be adjusted for the reserve margin and losses.

In our simple case of having only two periods, peak and off-peak, all related capacity costs are charged to peak consumers. However, if outage costs are expected to be significant outside the peak we allocate marginal capacity costs over several tariff periods, according to some heuristic rule. Section 9.6, taken from the World Bank Staff Working Paper No. 340,[1] gives further illustration.

Ideally, all transmission and distribution investment costs are allocated to incremental capacity related costs, because the physical designs are basically determined by the peak kW that they carry and

not the kWh serviced. We introduce voltage level into the tariff
at this stage by using the capacity costs at each supply voltage.
The average incremental costs of transmission plant are annuitised
using the test rate of discount over the economic life of the plant
(say) 30 years. This provides the transmission element of marginal
costs to add to the generation element, taking account of losses.
This procedure may be repeated at all levels of transmission and
distribution.

The LRMC of transmission and distribution is based on actual growth
of future demand, and averaged over many consumers. As we shall see
in Chapter 10, there are some exceptions to this averaging. Firstly,
some transmission is specifically associated with connecting to the
grid particular generating stations, e.g. hydro stations. The cost
of such transmission lines is normally considered as a generation
cost and not as a transmission cost. Secondly, transmission is
sometimes associated with a particular load area; the capital costs
are then allocated accordingly to the loads in that area. Thirdly,
some transmission can be identified for specific consumers. The costs
of such transmission are often allocated as consumer-related costs;
e.g. the transmission to a cement works is considered to be
recoverable directly through a 'consumer contribution'.

Similarly it is important to divide the distribution investment
costs between low voltage capacity-related costs and consumer-
related costs. A customer with a long low-voltage line should have
the line cost specifically allocated to that consumer as a capacity-
related cost. It should not be included in the averaged incremental
capacity-related costs for low-voltage networks.

9.5.4 Marginal energy-related costs

The LRMC of energy during the peak period will be the running costs
of the generators at the most costly end of the merit order. In Fig.
9.8 these meet the incremental peak kWh at ΔD. Under our assumption
of only a peak and an off-peak period this is the fuel and operating
costs of gas turbines adjusted by the appropriate peak loss factors
at each voltage level, as in the case of marginal capacity-related
costs.

The LRMC of our off-peak energy period corresponds to a load
increment during the off-peak period. This is usually the running
costs of the least efficient base-load or mid-merit plant used during
this period, suitably adjusted for loss factors, which will be smaller
than for the peak period in that I^2R losses are a function of the square
of the current I, and the current I is greatest during the peak period.

A predominantly hydro power system may be energy constrained. It
may be short of storage (kWh) rather than generating capacity (kW).
Incremental capacity is then needed primarily not to supply more
capacity but rather to generate more energy. In this case the energy
shortage constraint comes into force before the capacity constraint
and there is little distinction between peak and off-peak costs, and

between capacity-related and energy-related costs. Hydro energy consumed during any period leads to an equivalent fall in level of the reservoirs, and it is sufficient to levy only a simple energy-related (kWh) charge at all times, e.g. by applying the AIC method of total incremental system costs.

Any time that hydro is involved, marginal energy costs must be determined carefully. We deal with this subject also in Chapter 8. At times when water is being spilled or run off for purposes of irrigation, the marginal energy-related costs are zero. However, when hydro is displacing thermal plant, the running cost of the latter is the relevant marginal energy-related costs. When the pattern of hydro and gas turbine peaking operating is changing rapidly from one to the other, the marginal energy-related cost is a weighted average, with the weights depending on the share of future generation by the different types of plant mentioned.

9.5.5 Consumer-related costs

Fixed consumer costs consist of non-recurrent expenses attributable to items such as consumer connections, meters and labour for installation. These costs are usually either charged to the customer as a lump sum or are charged as distributed payments over several years. Recurrent consumer costs occur due to meter reading, billing and administrative expenses. Consumer-related costs are often imposed as a flat charge, in addition to the capacity-related and the energy-related charges. These costs are usually very small.

9.6   APPENDIX B: EXTRACT FROM WORLD BANK STAFF WORKING PAPER 340*

9.6.1 Allocation of capacity and energy costs among peak and off-peak users

The simplified model of a typical electric power generation system is used below to show from a conceptual viewpoint, how a long run marginal cost (LRMC) analysis based on the optimal system expansion plan yields the following idealised conclusions.[14]

(1)   Peak users should pay off-peak LRMC of capacity as well as energy costs.

(2)   Off-peak users should pay only off-peak LRMC of energy.

(3)   LRMC of peak capacity = LRMC of base load capacity - net fuel savings due to this base load plant.

---

* Extracted by permission from the World Bank

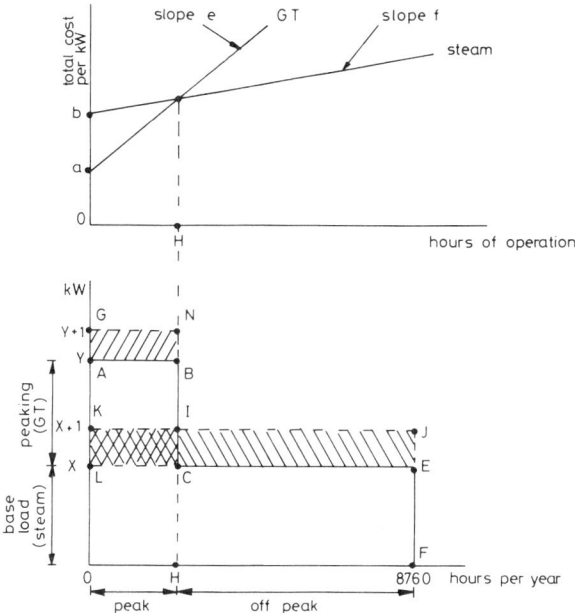

Fig. 9.10 Plant costs and annual load duration curve (LDC)

World Bank Staff Working Paper 340. Reproduced by permission of the World Bank

Consider an all thermal generation system having the annual load duration curve (LDC) shown in Fig. 9.10. There are only two types of plant whose linearised cost characteristics are given in the Table below, and also in the Figure. We ignore for the moment, all losses, reserve margin, etc.

Plant type.	Capacity cost per kW installed (annuitised)	Operating costs per hour
1. Peaking (e.g gas turbines: GT)	a	e
2. Base load (e.g. steam)	b	f

Total cost of 1 kW which is used h hours per year:

GT   : $a + e.h$

Base : $b + f.h$

Let H be the hours of operation which corresponds to the crossover point for which GT and base-load plant total costs are equal. Therefore, $a + e.H = b + f.H$

$$H = \frac{b-a}{(e-f)} \qquad\qquad (9.1)$$

The most economic use of plant can be determined by examining the LDC, OABCEF:

(1) For planned base-load operation (i.e. more than H hours per year), use base load plant; X kilowatt
(2) For planned peak operation (i.e. less than H hours per year), use GT; (Y-X) kilowatt.

Total annual costs of meeting the demand depicted by the LDC is:

$$C_0 = X(b + f.T) + (Y - X)(a + e.H)$$

Case 1: Only peak period demand increases by 1 kW (as shown by shaded area AGNB in Fig. 9.10). The optimal system planning response is to increase GT by 1 kW.

Total annual cost is $C_1 = X(b + f.t) + (Y + 1 - X)(a + e.H)$

Therefore, increase in cost:

$$C_1 - C_0 = a + e.H$$

This is the increase in system costs incurred because of the 1 kW increase in marginal (or incremental) demand during the peak period, and thus the peak period user must pay this cost.

The peak costs consist of:

(1) Capacity charge = a per kW per year
(2) Energy charge = e per kWh.

It may be seen that peak users' payment = $a + e.H$ = increase in sytem costs.

Case 2: Only off-peak period demand increases by 1 kW (as shown by shaded area CIJE in Fig. 9.10).

The optimal system planning response is to add 1 kW more of base load plant. But now there is 1 kW less of GT required than before.

Total annual cost: $C_2 = (X+1).(b + f.T) + [Y-(X+1)] (a+e.H)$

Therefore, increase in cost:

$C_2 - C_0 = (b+f.T) - (a+e.H) = (b-a) + (f-e).H + f(T-H)$

Substituting for H from eqn. 9.1

$C_2 - C_0 = (b-a) + (f-e).(b-a)/(e-f) + f.(T-H)$

$C_2 - C_0 = f(T-H)$

Therefore, the increase in system cost incurred due to the 1 kW increase in marginal off-peak demand is equal to the energy cost of operating the base-load plant during this period, and thus off-peak users must pay only this energy charge f per kWh. There are no capacity costs incurred by off-peak users, since:

off-peak users' payment = $f(T-H)$ = increase in system cost

In particular, we note that the base-load capacity cost b is exactly offset by the total cost saving due to GT, which is not required any more (i.e. capacity cost a plus net fuel cost saving $(e-f).H$ inside the shaded area LKIC). In other words: peak capacity cost = base-load capacity costs - net fuel savings:

$a = b - (e-f).H$

Case 3: Both peak and off-peak demand increases by 1 kW. This case is a linear combination of Cases 1 and 2, and therefore consumer charges are:

(1) Peak capacity charge: a per kW per year;
(2) Peak energy charge: e per kWh;
(3) Off-peak energy charge: f per kWh.

Clearly, total peak and off-peak users' payment = $b+f.T$ = increase in system cost. These results may be generalised to include more types of generating units, and rating time periods; e.g. n plant types and n rating periods, where these rating periods are chosen to coincide with the economic crossover points between competing types of plant.

In this case LRMC prices would be:

0 to H_1 = peak period: capacity charge a_1 per kW per year, and energy charge e_1 per kWh

H_1 to H_2 = 2nd period: only energy charge e_2 per kWh

H_{n-1} to T = nth period: only energy charge e_n per kWh.

Note: A more realistic system model would have to consider a number of complicating factors such as a larger number of rating periods and plant types, non-coincidence of the rating periods with the economic crossover points between different plant types, economies of scale and variable heat rates for a given plant type, hydroelectric plant including pumped storage, reserve margins and stochasticity of supply and demand, and so on. The most important difference with respect to the general case is that some capacity costs would have to be borne by consumers outside the peak period, However, the bulk of the capacity costs would still be allocated to peak period users. A simplified exposition of this result is provided in Reference[14].

REFERENCES

[1] TURVEY, R., and ANDERSON, D.: 'Electricity economics'. Published for the World Bank (Johns Hopkins University Press, Baltimore, 1977); also CICCHETTI, C.J., GILLEN, W.J., and SMOLENSKY, P.: 'The marginal cost and pricing of electricity' (Ballinger Pub. Co., Cambridge, Mass., 1977); also MUNASINGHE, M.: 'Marginal cost based tariff calculations in developing countries'. EWT Dept. Report, World Bank, Washington DC, 1979; also 'Electric Power Pricing Policy'. World Bank Staff Working Paper No. 340, 1980, (this has been drawn upon by the author for Appendix 9.5 and Appendix 9.6 of this Chapter by permission of the World Bank); also BAUMOL, W.J., and BRADFORD, D.J.: 'Optimal departures from marginal cost pricing', Am. Econom. Rev., June 1970, pp. 265-283; also FELDSTEIN, M.S.: Distributional equity and the optimal structure of public prices', Am. Econom. Rev., 1973, pp. 32-36

[2] A good survey of the literature on time-of-day, etc. pricing and load management on power systems is given in : MORGAN, M.G., and TALAKDAR, S.N.: 'Electric power load management: Some technical, economic, regulatory and social issues'. Proc. IEEE, 1979, 67, (2), pp. 241-313. Other useful papers are given in References 3, 4 and 5

[3] ACTON, J.P., GELBARD, E.H., HOSEK, J.R., and MCKAY, D.J.: 'British industrial response to the peak load pricing of electricity'. RAND report R-2508-DOE/DWP, Santa Monica, California, Feb. 1980

[4] MANICHAIKUL, Y., and SCHWEPPE, F.C.: 'Physical/economic analysis of industrial demand'. IEEE Trans., March/April 1980, PAS-99, (2), pp. 582-588

[5] MIT Energy Laboratory: ' New electric utility management and control systems', Proceedings of Conference, MIT Center for Energy Policy Research and Electric Power Systems Engineering Laboratory, Report EL 79-024, June 1979

[6] BOHN, R.E., CARAMANIS, M.C., and SCHWEPPE, F.C.: 'Optimal spot pricing of electricity: Theory'. MIT Energy Laboratory Working Paper, MIT-EL 81-008WP, March 1981; also 'New electric utility management and control systems'; Proceedings of Conference held in Roxborough, Mass. 30 May-1 June 1979, by the Homeostatic Control Study Group, MIT Energy Technical Report, MIT-EL-79-024

[7] 'Trials for Seeboard's Interactive Control', Elect. Rev., 26 June 1981

[8] CAMM, F.: 'Industrial use of cogeneration under marginal cost electricity pricing in Sweden'. RAND Report WD-827-EPRI, 1980; also BORN, R.: 'Industrial response to spot electricity prices: some empirical evidence'. MIT Energy Laboratory Working Paper, MIT-EL-080-016WO, Feb. 1980; also GORZELNIK, E.F.: 'T-O-U rates cut peak 3.5%; kWh by 1.3%. Elect. World, 15 Sept. 1979, pp. 138-139

[9] MITCHELL, A., MANNING, W.G., and ACTON, J.P.: 'Peak load pricing: European lessons for US energy policy' (Ballinger Pub. Co., Cambridge, 1978)

[10] ICF, Inc.: 'Interruptible electric service for industrial and large power customers'. Washington DC, Unpublished, May 1980

[11] SCHWEPPE, F.C.: 'Power systems 2000', IEEE Spectrum, July 1978; also SCHWEPPE, F.C., TABORS, R.D., KIRTLEY, J.L., et al: 'Homeostatic utility control', IEEE Trans., May/June 1980, PAS-99, (3); also KEPNER, J., and REIGNBERGS, M.: 'Pricing policies for reliability and investment in electricity supply as an alternative to traditional reserve margins and shortage cost estimations'. Unpublished paper, 1980

[12] BALERIAUX, E.J., and DE GUERTECHIN, L.E.: 'Simulation de l'exploitation d'un parc de machines thermiques de production d'électricité couplés à des stations de pumage', Revue E, (edn. SRBE), 1967, 5, (7)

[13] FINER, S.: 'Electric power system production costing and reliability analysis including hydroelectric, storage and time dependent power plants'. MIT Energy Laboratory Technical Report, MIT-EL-006, Feb. 1979

[14] WENDERS, J.T.: 'Peak load pricing in the electric utility industry', Bell J. Econom., Spring 1976, 7, pp. 232-241

Chapter 10

Network economics

10.1 INTRODUCING GEOGRAPHY

10.1.1 The problem

Until this Chapter we have basically assumed a single node system. Although it is customary to start system planning with this assumption we must find a way of proceeding from a single to a multi-node system.

Theoretically we must consider every alternative combination of generation plus transmission to meet a given load forecast at an optimal reliability standard. The combination with the lowest total present value of system plus outage costs should be chosen.

This is an immense task for all but the smallest systems. There are 2 to the power of $n(n-1)/2$ combinations for the number of ways of connecting up n nodes if all ways of making combinations are counted.

Number of nodes	Number of ways of connecting the nodes
1	1
2	2
3	8
4	64
5	1024
6	32768

The system planner soon loses direct comprehension above three nodes although a computer can handle more. Planning the British 132kV and 400kV grids was done on a 2-node basis. Semi-arbitrary boundaries were given to geographical areas and the transmission strength between any two areas chosen by (i) an empirical formula derived from past experience of the maximum amount of power flow needed from one area to cope with plant outages in the other; plus (ii) planned power transfer across the boundary.[1] By this means we can divide an interconnected system into a number of 2-area 'sets' and find the transmission requirement for each set, thus building up a network to satisfy all the transmission requirements; the more sets there are the better our network will be.

The process can be repeated for any number of years into the future; a dynamic network will emerge to cope with various options for sitings, size and mixes for generation. Given sufficient empirical data this method has much to recommend it.

The above two-node approach can be used most easily for a large interconnected system for which historic data are readily available, and especially for a uniformly loaded network which needs reinforcing at a higher voltage.

10.1.2 Multi-nodal systems

Fortunately in practice it is usually possible to single out the dominant generation and loading nodes. It is then merely a question of deciding upon the various actual combinations of generation/load interconnections needed to be examined to determine that combination with the lowest total present value of system plus outage costs. Optimising the generation plus network is carried out by the normal process of marginal analysis of the background plan(s) (Chapter 8).

It is normal to reduce the complexity by prescribing against particular interconnections. This simplifies the problem; for example a 5-node system:

Number of interconnections prescribed against	Combinations of interconnections	Number of lines
0	1024	10
1	512	9
2	256	8
3	128	7
4	64	6
5	32	5
6	16	4
7	8	3
8	4	2
9	2	1

With a large interconnected system, by this means it is possible to reduce the vaious generation/network combinations to a manageable number.

10.2 FUNCTIONS OF NETWORKS

10.2.1 Introduction

To find the optimum generation/network structure it is also necessary to have some philosophy for network structure formation.

The following questions have to be asked.

(i)   Should generation be interconnected, regardless of load centre geography, e.g. so that spare generation can be optimally pooled and the highest in merit-order always used?

(ii)   Should load centres be interconnected, regardless of generation geography, e.g. for greater security, increased ability to re-establish supply, smaller system losses, etc.?

(iii) Should a generation-interconnection plan take precedence over a load-centre-interconnection plan, if the two plans cannot be reconciled?

(iv)   How much interconnection should we have, e.g. in terms of transmission capacity? A massive interconnected system is costly but has a high level of security and enables new plants and new load centres to be easily connected. A fragile interconnected system is less costly in capital, losses and operating equipment, e.g. national and local control centres, but its level of security is low. This problem can only be resolved by optimising system plus outage costs (Chapter 3).

(v)   How many tiers of voltage should there be and should they be operated electrically as a single system?

These questions are partly academic except where a green-field situation exists; the shape of a network is usually irreversible beyond a certain point in the development programme because it would be uneconomic to change it. However, in most developing countries and whenever a further layer of voltage is to be introduced the pattern has not yet been set and the above questions are very pertinent.

10.2.2 Planned transfer function of networks

The planned transfer function concerns transferring energy between two different geographical points in the form of electricity. For the function to be economic it must be compared with transferring the same useful energy in the form of fuel for generation at the load centre. The economic viability is tested by calculating the total attributable annual cost, capital and running, over the economic lives of the equipment, for each method for the same energy transfer. Attributable capital costs are represented by annuities calculated over the economic life of the equipment using the normal discount rate as the rate of interest.

A common case in the developing world is to evaluate the economics of the transference of particular hydro energy in bulk. The direct comparators for the particular hydro plus attributable network are:

(i)   other hydro generation plus attributable network;
(ii) thermal generation plus attributable network; or
(iii) thermal plant near the load centre plus fuel transport.

The same amount of useful energy is assumed in each case.

The above network economics are relatively simple because the same amounts of energy are assumed. If dissimilar amounts of energy are being considered then the more complex methods described in Chapter 8 must be used, i.e. the network must be considered as part of the whole system contributing present value of capital and running costs.

10.2.3 Capacity pooling function of networks

If two or more generating plants are interconnected, then each plant can stand by for other plants during outages. The savings in generation reserve capacity, i.e. the planning plant margin in planning terms, and in spinning reserve, hot-standby and cold-standby generating plant in operational terms, can thus be offset against the capital and running cost of the interconnection. To calculate the economic worth of this function we must model the performance of the whole system, with and without interconnection, to determine whether the total present value, over the economic life of the interconnection, of the savings in generation reserve is greater than the capital cost of the interconnection; if so the interconnection is justified.

In practice there seems to be a 'threshold' of interconnection capacity before which it is difficult to justify any interconnection, but beyond which additional interconnection can be easily justified. In a marginal case the final decision to interconnect may be made for other reasons, e.g. greater operational flexibility, greater system security. It may be possible to justify only 'partial' interconnection, e.g. by using standby circuits operated 'normally-open'.

10.2.4 Economic-operation function of networks

A fully interconnected system allows maximum possible savings in total system running costs by incorporating all generating plants into one merit-order list. The savings in system (mainly) fuel costs with and without any interconnection can be set against the capital and running cost of that interconnection. The methods of system modelling are again used to determine the economic worth of this function.

10.2.5 Networks acting as multi-purpose projects

Because networks usually have more than one function many planners believe that the economic methods applicable to multi-purpose projects[2] should be used. To the author's knowledge this has never been done.

10.2.6 Other functions of networks

Functions of networks whose costs and benefits may be more difficult to quantify than those above are:

(i)   The ability of a 'grid' to do away with geography, i.e. to enable power stations and large loads to be sited with little reference to network cost.

(ii)   The function of high power distribution, i.e. supplying large high-density loads in metropolitan and highly industrialised areas.

(iii)   International interconnection and inter-power-system inter-connection. For example, the 'tie-line' connections already common in large continental areas, e.g. North America, USSR and parts of Europe. These tie-line systems are now becoming important in developing countries. Tie-lines are used for closely defined power transfers, often only for emergency use, which is usually smaller than transfers on normal inter-connections.

(iv)   Interconnection enables generators of larger capacity to be introduced earlier. This is partly an extension to the economic-operation function described above.

10.3 ECONOMIC JUSTIFICATION OF A NETWORK[3]

10.3.1 Finding a least-cost solution

Assuming a load forecast for daily, monthly and annual maximum demands and energy requirements of the system(s) and taking geography into account, the least-cost solution is that which meets the load forecast at a minimum total present value of system plus outage cost over the economic life of the interconnection project.

The simplest comparison is of different generating stations, plus associated interconnection, for supplying the same demand for electricity against time at a given standard of reliability for all years considered.

The investment costs of each generating station, plus associated interconnection, are annuitised over the station economic lives using a sinking-fund type of depreciation at a rate of interest equal to the discount rate. These annuitised annual costs can then be expressed in terms of annual charges (cents/kWh) of electrical energy delivered.

To get the total cents/kWh delivered, the fuel and other operating costs per kWh of the options must be added to the annual charges after allowing for losses. Planned transfer interconnection schemes can then be compared directly on a total cents/kWh delivered basis.

With more than three planned transfer options, possibly operating at different load factors, the total cents/kWh method becomes very difficult and the more complex methods of Chapter 8 have to be employed, these methods using discounted cash flow techniques:

(i)   identify investment, running and outage costs for each network option being compared in each year of the option's economic life
(ii)   discount these annual costs using an agreed rate of discount
(iii) sum algebraically the discounted annual costs to a 'total present cost' of each network option and choose the option with the lowest total present cost.

Constant price levels are used, without taxes and subsidies, with shadow pricing as appropriate to correct for distortions in the economy due to artificial rates of foreign exchange, labour, capital and fuel.

10.3.2   Economic return

The economic return (ER) determines whether any option at all should be built. Perhaps the investment should be made in another sector?

The major difficulty in calculating the ER lies in attributing benefits. The total annual kWh attributable in any year to the incremental network purchased by the investment is a first measure of the benefit of that network in that year. Turned into cash terms by multiplying the attributable kWh by the tariff, these benefits form a cash flow over the economic life of the network to match the equivalent cash-flow investment running costs and outage costs. Benefits to the economy are added; e.g. contribution to economic growth, environmental benefits.

The economic return is the discount rate which equates the total present values of attributable costs and attributable benefits over the economic life of the network. Only projects with an economic return greater than the opportunity cost of capital are acceptable.

10.4   TRANSMISSION

10.4.1   Introduction

There is no clear dividing line between transmission and distri-bution. In general, transmission is concerned with the movement of

electrical energy on a much larger scale than in distribution. A
good illustration of the framework of modern transmission systems
is given in Figs. 10.1 and 10.2, dealing with Europe and South-East
Asia. The problem is to find the optimum network structure on a
continental scale bearing in mind all the functions of networks
mentioned above.[4]

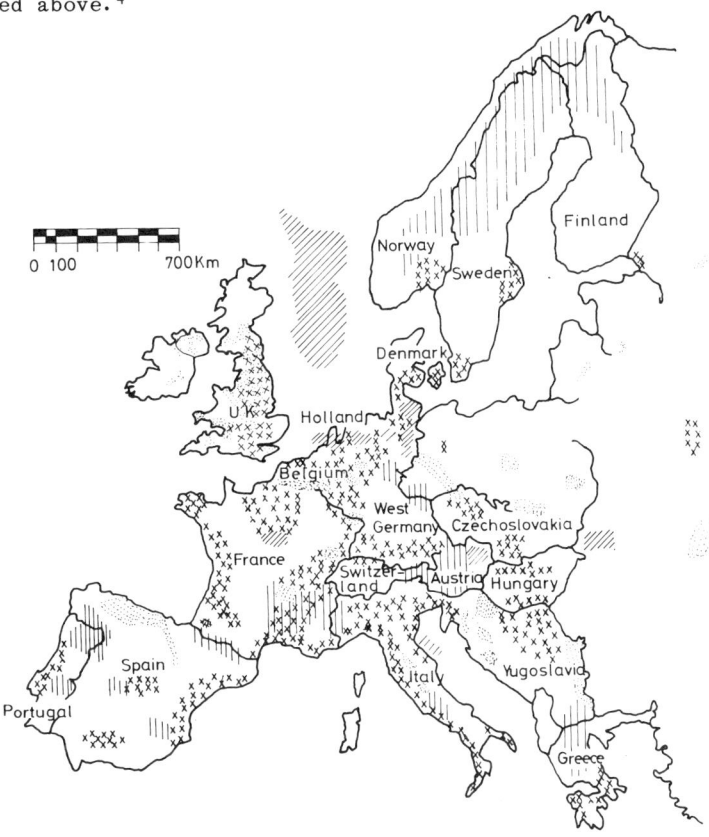

Fig. 10.1   Energy statistics for Western Europe

⊠⊠⊠ Population over 50 per km^2

▥▥▥ Hydro Potential

▨▨▨ Oil Fields

▦▦▦ Coal and Lignite

UN Statistics, OECD Statistics and Electricity Authorities' Statistics and Reference[5]

Fig. 10.2   South-East Asia: Main population areas and resources for
power generation

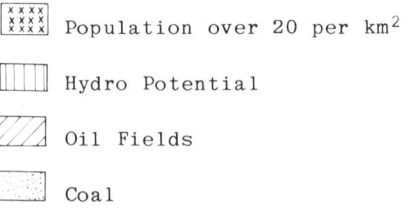   Population over 20 per km^2

Hydro Potential

Oil Fields

Coal

BERRIE, T.W.: 'Prospects for interconnection in South-
East Asia'.   Elect. Rev., 7 Dec. 1979

Point to point transmission in the UK began in 1835 when Deptford
Power Station supplied the Thames Embankment and Grosvenor Galleries
at 10 kW. The 1920s and 1930s saw the construction of the UK 132kV
'grid iron' system which reduced the amount and cost of generation
plant required. In 1927 power was produced in the UK at 0.5 old pence
(US¢0.4 per kWh) when domestic coal cost about £2.50 per ton
delivered. Electricity has maintained a relatively constant relationship to fuel prices in the UK ever since, partly due to a fuel-cost
adjustment clause in tariffs in more recent years. Today the UK,
Europe and Japan have large heavily loaded 400 kV 'super-grid'
networks using the previous transmission 'grids' as primary
distribution.[5]

Either AC or DC current can be used for long-distance networks.
With respect to network voltage levels, 1100 kV AC is now appearing
in the United States, and other countries have 800 kV networks.

Zaborsky and Rittenhouse produced one of the best network design
reference books in the 1950s.[6] It covered all factors determining
power flow in networks; 'capability charts' with MW limits were set
by an envelope of voltage drops, phase angles and thermal constraints.
A cruder set of charts to do the same job is given in Figs. 10.3-10.6.
As the load carried by a circuit increases beyond the set limits the
next higher voltage is chosen. This has led to the level 110kV to
150kV being the 'first' network voltage level, followed by discrete
220kV-330kV being overlaid at the 'second' voltage level. These in
turn are being overlaid at a 'third' level of 380kV-500kV networks
and, at the 'fourth' level 750kV-1100kV networks are now being needed
in some areas to interconnect the main centres of large concentrations
of either load or of generation, using either AC or DC. Load sharing
is not easy over interconnections when a network is at different
voltage levels. Thus the highest last-adopted system interconnected
voltage takes over the network or 'grid' role, leaving the previous
interconnected voltages to act as sub-transmission or primary
distribution, as we saw earlier, the UK 132kV system took on the role
of primary distribution.

There are two frequency standards, i.e. 50 and 60 Hz. Roughly
speaking, North America, the Philippines, Saudi Arabia, Taiwan and
both Koreas use 60 Hz; the rest of the world uses 50 Hz, while Japan
has both frequencies with DC interconnection, possibly an ideal way
of operating systems with different frequencies side by side.

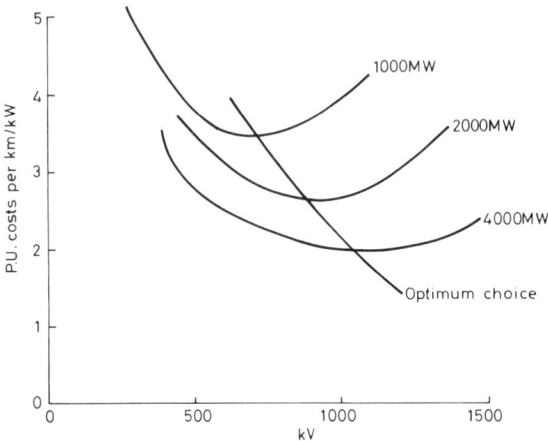

Fig. 10.3 Significant minimum costs of transmission including losses*

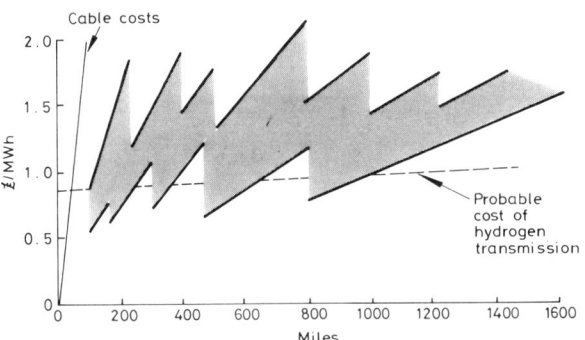

Fig. 10.4 Overhead lines - Costs of energy transmission*

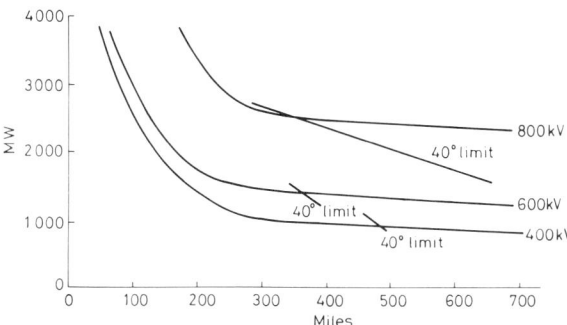

Fig. 10.5   Received power capability; for 10% regulation, unity power factor load for 400, 600 and 800 kV, 40^0 transmission angle*

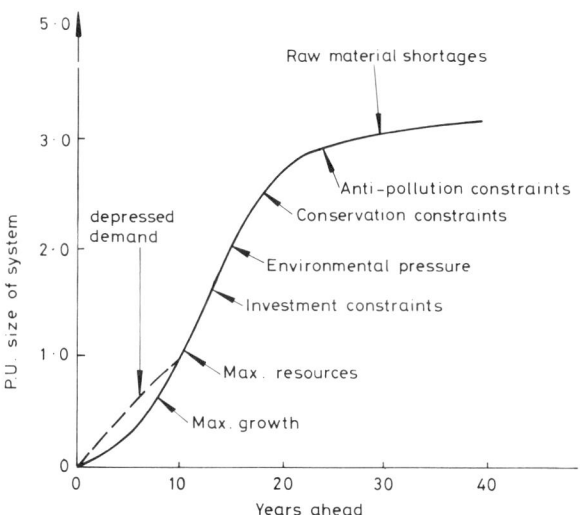

Fig. 10.6 Size of system*

*EGGLETON, M.N.: 'Scenarios of network structure'.   Paper for International Symposium on Electricity Economics and Load Management, Imperial College, London, 20-24 March 1980.   Reproduced by permission of Imperial College, Power Systems Group

10.4.2 Major networks

A summary of the world's five major networks is given below (mid-1981):

USA	(Loose ties) about 2500 TWh from about 650 GW
USSR	(Weak systems) about 1300 TWh from about 270GW
Western Europe UCPTE[7]	(Strong ties) 1700 TWh from about 570 GW
Eastern Europe	(Weak systems) 420 TWh from 87 GW
UK	(Strong systems) 300 TWh from about 75 GW

The world's electrical energy consumption over recent years is as follows:

1972	5644 TWh
1975	6462 TWh
1978	7300 TWh from 1830 GW
1981	8250 TWh from 2070 GW (author's estimate)

$$(1 \text{ TWh is } 10^9 \text{ kWh})$$

Of the five great power systems, the USA has the highest load factor with a summer rather than the winter peak of the other four. The UCPTE system has the highest power density of the five. However, whilst all five major systems are in contiguous network form, they represent very different structures. The USA is supplied from nine power pools and four adjacent Canadian systems. Its power exchanges between utilities are tightly controlled for economic and security reasons; power transfers are limited by controls co-ordinated to generators on 'tie-line' control. The power pools are designed for self-sufficiency under normal conditions.

The western part of the USSR system is basically similar to the USA except that DC is used to interconnect some control zones. Only very frail ties exist to remote eastern areas from the west. Disconnections of complete zones can occur. Some regions, however, remain interconnected after a period of asynchronous operation, a phenomenon not readily tolerated on other systems of the world.

Western Europe has a system interconnecting nine countries. Power exchanges between countries are by tie-line operation as are those for Eastern European systems which are being interconnected to Western Europe by DC links.

Of the five major systems only the UK system is under unified control. It operates to two national control centres, one in England and the other in Scotland. Japan has two systems interconnected by DC. The sector is made up of many supply companies contributing to a national power pool.

As well as geography, an important rôle played in the development of networks is the density of area load and generation. Network development obeys the following type of trend, taken from the England

and Wales system:

Example	Total load	Load density	Type of network
Central London	2000MW	100MW/km^2	275kV cables
Greater London	10000MW	20MW/km^2	400/275kV
Southern England	30000MW	1MW/km^2	400/132kV

The density in most developed countries is less than 0.5 MW/km^2.

China's emerging large system with about 60 GW of generation is scattered over large areas with insufficient transmission capacity to support adjacent areas. India's system is also large, with over 25 GW of generation.

10.4.3 Likely developments of existing networks

Estimates of likely developments of network structures are (Reference[4]):

UK
400kV, 275kV, 132kV   No higher than 400kV required before 2000, if at all

Italy
420kV, 145kV   1000kV not required until after 1990

Norway
380kV, 275kV   800kV link possible with Sweden

Australia
500kV, 330kV, 220kV   No higher voltage required until beyond 1990, 750kV being considered

Finland
400kV, 220kV   No higher voltage required until 2000, if then

USA
765kV, 500kV, 345kV, 138kV   Future possibility of 1100/1200kV
±400kV all DC   with 1500kV or above as an overlay

Germany, Federal Republic of
350kV, 220kV, 110kV   No higher voltage until beyond 1995. Later 800kV/1200kV overlay could be used

South Africa
400kV, 275kV   760kV being considered for the future, but nothing higher than 400kV required before 2000

USSR
750kV, 500kV, 330kV, 220kV   Future 1150kV and 1500kV likely, using
110kV, ±400 DC   AC

Belgium
400kV, 220kV, 150kV   No higher than 400kV required before 2000

Canada (Ontario Hydro)
500kV, 230kV   No higher than 500kV required until
                                     after the late 1990's

Brazil
500kV, 440kV, 345kV, 230kV,   750kV AC and ± 600kV DC being used
150kV   from Itaipa hydro station DC inter-
                                     connection with Uraguay and Paraguay

Sweden
380kV   800kV possible in future

Venezuela
400kV, 230kV   Considering 800kV

Czechoslovakia
400kV, 230kV   850kV or above required only after
                                     2000, but initially earlier for
                                     interconnections with Central Euro-
                                     pean elecricity systems; it has
                                     already been decided that this will
                                     be DC

Saudi Arabia
500kV, 380kV, 230kV   500kV being built

Mexico
400kV, 220kV   A higher voltage is needed

To economically justify such transmission systems we must show that
they are the generation-cum-network alternative with the lowest net
present cost (see earlier in this chapter).

10.5   DISTRIBUTION

10.5.1   Economic justification

Distribution economics in any formal sense is relatively new.
However, because distribution investment may amount to 40% of total
investment in the power sector we must satisfy ourselves that it is
just as cost effective as generation or transmission investment.
It is not always possible to distinguish between the functions of
transmission and distribution, especially in systems of smaller
developing countries where loads are often fed from single generating
stations, or at most from two or three interconnected stations; many
of the functions of transmission described earlier are then carried
out by distribution.

Often, because of its diffuse physical nature we cannot test the
economics of distribution systems as rigorously as we can the
economics of generation-cum-transmission.   We must proceed more
obliquely:

Firstly,   we check that the proportion of distribution investment
           in the investment programme for the whole power system is
           in line with past trends; in a 'green-field' situation we
           rely on data for similar areas or countries, by population
           density, agriculture, geography etc.

Secondly, when parts of the distribution system have a particular
function e.g. supplying a new town on a new industrial
estate, it is possible to attribute costs and revenues to
a particular physical group of distributors. In such
circumstances it is customary in determining attributable
costs to take the generation cost as either the bulk supply
tariff or an estimate of the long-run marginal cost of
supply which cost includes any attributable transmission
costs. The attributable costs and benefits enable an
economic return on the distribution investment to be
determined. This can be done for different supply
arrangements, including autogeneration, to see which has
the greatest return. However, if the return is low, before
rejecting the project altogether, we must ask whether the
low return is due more to the tariff level being too low
than the project not being worth while. Also, possibly
there are secondary benefits not represented by the
revenues which we should try to evaluate before rejecting
any of the schemes. Normally the economic return should
equal the opportunity cost of capital.

Thirdly, as for all networks, the main economic test of distribution
is to assume that the benefits are equal for all alternative
schemes and to find the least-cost solution for meeting
a given load growth (see earlier for transmission).

Fourthly, as with transmission, it is often more practical to deal
with distribution under its various functions, as we now
examine.

10.5.2 System reinforcement function

System reinforcement of distribution is more likely than that of
transmission to require an extra tier of voltage. As we have already
noted, this produces a new 'green-field' situation. The next voltage
tier might be difficult to justify formally economically because of
the large amount of 'lumpy' investment needed on which no quick return
is likely; the decision to install a higher voltage might well then
be taken on non-economic grounds, e.g. required flexibility of system
operation, difficulty of obtaining wayleaves for further
distribution routes.

10.5.3 System rehabilitation function

When capital is restricted distribution systems are the first to
be affected by cut-backs, partly because of the long time generation
and transmission projects take to build and partly because capital-
expenditure cuts on distribution systems affect only a small number
of geographically separated consumers. For the above reasons many

distribution systems tend to become sub-standard with time, unless a special watch is kept; in many developing countries an extra investment in distribution 'rehabilitation' is required at regular intervals for this reason. In practice, it is difficult to separate this distribution function from the function to be discussed next.

10.5.4 System replacement function

Shutting down old generation is a normal alternative to be borne in mind when deciding on the size and composition of generating development programmes. No similar consideration is given to the replacement of networks even though distribution in all major cities in the world must be nearing the end of its physical life, say 50 years, and even though economic calculations concerning the selection of a distribution scheme explicitly attribute an economic life to the distribution equipment. In order to restore true investment perspective it is necessary to determine whether it is economic or not to replace some distribution equipment now as an option when determining the least-cost distribution development programme.

10.5.5 Distribution reliability

In recent years the security standard to which urban distribution networks should be planned has become of increasing importance, since more and more system planners have realised that a change in distribution standards could lead to large savings in up to 40% of total capital investment. In the past, urban electricity distribution systems in developing countries were largely based on the reliability standards used in developed countries; many now consider that this approach spends unwarranted amounts of capital investment. Lending agencies are trying to insist that distribution reliability standards are worked out on an economic basis in the same way as standards for generation and transmission are calculated.[8]

This means that the quality of supply should be set at that level where an incremental investment in distribution just equals the incremental outage costs to consumers and the economy attributable to the investment, measured through the potential cost to consumers and the the economy of loss of supply at that reliability standard (see Chapter 3). This test is beginning to be obligatory for most urban distribution projects in developing countries sponsored by such lending agencies. The test is often difficult to do, although a good deal of work on it has been done by research programmes, e.g. that sponsored by the World Bank, as given in Reference[8]. Meanwhile reliability of distribution systems is seen more in the context of the 'type' of system with which we are dealing, as we now examine.

10.5.6 Radial distribution systems

A distribution system at its simplest consists of a node or point

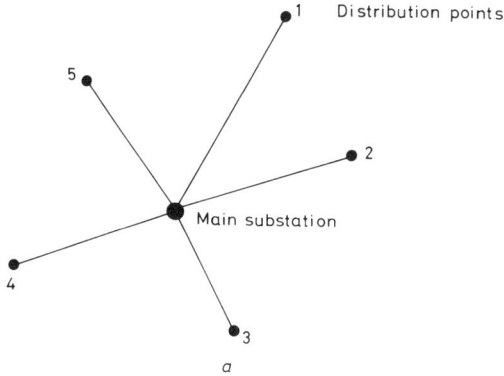

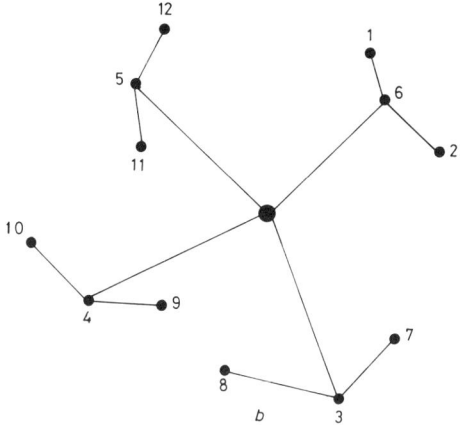

Fig. 10.7   Types of distribution system

 a Radial system
 b Radial distributor system

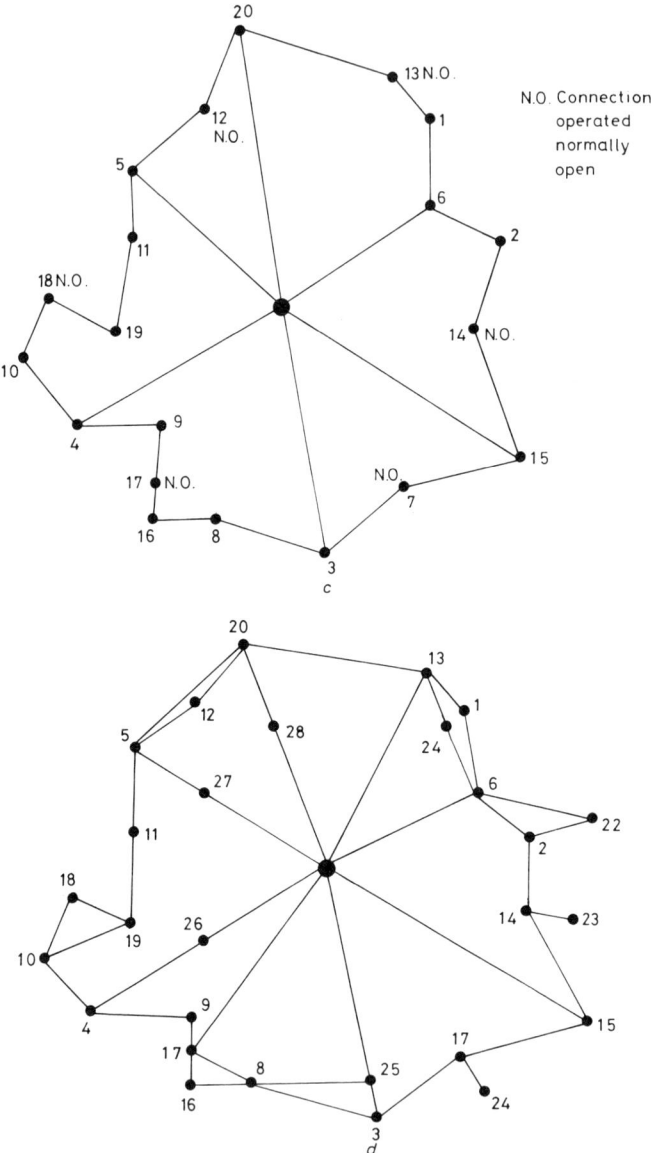

N.O. Connection
operated
normally
open

c   Ring system
d   Fully meshed system

of supply to which is connected a number of node-point loads. When
each of these point-loads has a separate connection to the source
of supply, the distribution system is described as 'radial'. It is
simple to identify the distribution costs of supply in a radial system
and to allocate common costs between loads in accordance with the
contribution which each load makes to the simultaneous or marginal
maximum demand at the node. Because the actual capacity of each feeder
is matched exactly to the maximum demands of each load, no benefit
is obtained of any diversity between these loads until we reach the
point of common coupling at the node or input source. One advantage
of a radial system is that the failure of one feeder will interrupt
only supplies fed by that feeder (see Fig. 10.7a).

Frequently the distance between loads is less than that between
loads and supply source. It is then feasible to move the point of
coupling between loads away from the input source and nearer to the
loads by establishing what used to be called a 'main'. The so-called
'service' lines of the loads are then connected to one of the 'mains'
at their nearest point of access. The 'mains' are connected to the
power source at their nearest point to it, forming a 'radial
distributor' (see Fig. 10.7b). Radial distributors permit one to
exploit the diversity between the maximum demands of individual loads
with a consequent saving in total system capacity. Failures of radial
distributors still interrupt all supplies fed off the distributors.
In a radial system only the individual service-line costs to each
load are directly attributable to each load. All source and mains
costs are shared between loads by some weighting process which might
include marginal costs.

10.5.7 Ring systems

The probability of supply interruptions on a radial distributor
is often unacceptably high as measured by any type of cost-benefit
analysis, because of the amount of equipment common to all loads and
exposed to the same fault risks. In such circumstances it is
economically justified to provide a connection to an alternative
source of supply for at least some of the loads. Such a separate
connection should be as remote as possible from the first connection
and all associated loads should be situated between the two
independent sources.

Establishing a second supply often dramatically improves
operational flexibility under both normal and fault conditions.
However, because network complications can arise from paralleling
two supply sources, it is common to run 'ring' systems of this type
'normally open' at the mid-point (see Fig. 10.7c); the second feeder
facilities are only used during planned or emergency outages. If
few network complications arise the network can be run permanently
interconnected, i.e. we have a fully meshed network (see Fig. 10.7d).

10.5.8   Continuous network development[9]

As loads increase, distribution systems evolve from the simple radial feeder to the radial distributor, to the ringed and fully meshed network types. In general, the level of distribution voltage is decided by the magnitude and density of the loads rather than cut-and-dried economics.

Since the basic structures and evolution of both high- and low-voltage networks are similar, one can 'overlay' higher over lower voltage networks. One can also 'inject' power from one network into another as a means of reinforcing any part of the lower-voltage network. This can be done most economically by deciding in advance what load growth is likely to create a shortage of capacity, i.e. by making a detailed geographic load forecast working to a grid reference system of the type shown later. As load densities continue to rise above the capability of the lower voltage, we must put in the most economic reinforcement in terms of system and outage costs, usually by establishing an additional step-down point from the next highest voltage network, following which the lower voltage network feeder arrangements are rearranged. The most economic procedure is to establish a new feeder area made up of parts of the feeder areas previously included in those of surrounding step-down points; this allows the feeding area and the loading of the surrounding step-down points to be reduced. Spare capacity has then been reintroduced in the most economic way possible by transferring some through-loads previously fed via the lower-voltage network to the higher-voltage network, as well as catering for the additional load which initiated the need for reinforcement.

Thus, as load densities increase, existing feeder areas decrease in geographical size and new ones are established. In this way loading on the distribution system is kept within the same capability criteria as when the network was originally established.

In the past this permitted incremental investment in distribution to be matched almost directly to incremental load growth. Coupled with the inherently long life of most distribution equipment, this means that networks once established do not change fundamentally in structure throughout their lifetime. Thus decisions with far-reaching economic consequences are made at the outset, before the distribution network has been established, on choice of voltage and distributor capacities; any such planning must not be over-influenced by cost-benefit studies related only to the load forecasts for the short and medium term. Economic analysis of distribution must take into account long-term load forecasts and future load types which might be quite different from the initial ones. We must also have a measure of flexibility in the plan; future loads are never exactly where, when, or of the size expected at the outset.

The usual pattern of distribution growth means that the utilisation of capacity follows a 'sawtooth' shape with time, each tooth representing a major reinforcement of the type mentioned. The gradient at the 'front' of the sawtooth depends on the load growth within the distribution step-down feeding area, the height of the tooth depends on the point in time when overload of the network occurs in each stage of its development. At one point in time an increment of load growth can be accommodated by the small cost of an extra connection; a similar increment at another point in time will 'trigger' a need for wider reinforcement and possibly the establishment of a new step-down point. In the ultimate, minor increments of growth can trigger large investment expenditures involving the reinforcement of a whole new supporting higher-voltage network.

The greatest difficulty in carrying out economic assessments of meshed distribution networks for specific increments of load growth may be the practical problem of defining the specific attributable area of influence of any one distribution development project or investment. This applies not only in the 'vertical' structure of overlayed voltage networks installed one above the other as described earlier, but also in the original lower voltage networks themselves. The problem of supply area attribution is at its worst when we run a 'solid network', i.e. a fully meshed low-voltage network with all step-down points electrically paralleled. However, one considerable advantage of such an arrangement is that all electricity load is shared out evenly among all transformers at the step-down points, and this can make the economic sums easier to do for both attributable costs and benefits. If, as is normal, those transformers are of standard size, they will have a uniform level of utilisation. As the load grows the utilisation level of the transformers increases uniformly across all transformers. Ultimately all transformers will overload together; at this stage any small increments of load growth at any point in the system triggers the need to carry out a complete network reinforcement scheme, and it may be more economic to reinforce before this happens. Reinforcement usually involves breaking up the nicely balanced meshed network and restructuring it into sub-networks which are far from balanced; each sub-network will require large injections from the next voltage level to restore network utilisation to a manageable level.

Throughout the development of any network the normal rules of cost minimisation and acceptable cost-benefit analysis are the guidelines for decision making.

10.5.9 Economics of a distribution planning plant margin

Distribution systems described up to this point consist of distributors and transformers connecting high-voltage networks to lower-voltage networks. If all such equipments were infinitely reliable, supplies to all loads would be continuous. There would be no further need for expenditure beyond that directly necessary to connect the load to the source of supply.

Unfortunately distribution equipment is not sufficiently reliable to provide automatically the supply standard expected by most consumers.   General investment in a planning plant margin for distribution plant is as necessary as for generation plant.   Specific investment is needed for somewhat different functions than in the case of the generation plant margin, e.g.:

(i)   for any hardware installed, to limit the number of loads directly affected when a fault occurs

(ii)   for alternative paths between the source of supply and the loads by providing more 'ring mains' or duplicate distribution feeders.

Within any system it should be possible to:

(a)   identify costs associated with those components whose sole function is to secure supplies against failure

(b)   calculate the outage costs savings to consumers and the economy by thus safeguarding supplies

(c)   roughly determine the optimum quality of supply under the reliability rule of Chapter 3.

The level of investment in distribution systems to enhance the standard of supplies is an option available to any power utility, and its economics can be tested as in (a), (b) and (c) above.

There is presently no universally accepted method of comparing the reliability of distribution systems.   If we could do a complete cost-benefit analysis at every stage of distribution planning this would give us basic economic indicators of the type we have for generation reliability (Chapter 3).   Attempts have been made at partial economic indicators for distribution, e.g. capital cost per additional kilowatt of demand supplied.   However, such economic criteria have in practice proved inconsistent, or have failed to take account of significant variables.   Many distribution planners have abandoned hope of ever having basic economic indicators, and believe that the distribution case is too complex. Certainly the distribution case is complex because of the possibility of so many joint probabilities of failure. Any solutions found will almost certainly need to be based on long-term whole-system costs similar to those for generation planning, instead of on measuring the effectiveness of incremental projects taken one at a time.   Perhaps the pursuit of short-term over-simplistic solutions has hindered the search for long-term economic indicators. We thus come back full-circle to the ultimate need to disaggregate the marginal analysis of a long-term background plan right down to the level of distribution as well as generation/transmission.   In this age of computers this does not seem to be an untoward request.

Investment in power distribution networks in most countries now represents one of the most important national assets upon which the economic prosperity of the nation depends. For example, in the United Kingdom the present replacement costs of the distribution assets now stand at about £20,000 million.   The annual running cost is about

£300 million. It is upon this distribution that the huge generation and transmission system is ultimately dependent for meeting the vast proportion of the load. Yet the amount added for the annual cost of these distribution systems on the price still remains at about 10% of the total price for each kilowatt sold.

10.5.10 Need for more case studies

Most publications on reliability deal only briefly with distribution. Reliability analysis seems to be a prerogative of generation and transmission planning. One reason for this is that more studies are needed of the 'standard' distribution design practices of utilities. In particular, we must explicitly consider the distribution planning plant margin as viewed from the quality of service being planned. Each quality of supply has its own incremental cost and its own incremental benefits. Thus more consumers can be supplied for the same investment cost at a lower quality of supply for all other consumers. Conversely, the net benefits from an improvemnt in service to all other consumers have to be set against the net benefits that might otherwise have been provided by having additional consumers. Also, lower unit costs can mean increased investment is available to another class of consumer altogether. This is a familiar problem in other sectors, e.g. transport, where the net losses to one class of road user due to lower standards are offset by the gains to other users. Similar techniques for solving such problems can be applied to the power sector.

There is a great need for more case studies of the type carried out by the World Bank, described in Reference[8], the results of which are given in Tables 10.1 - 10.4 and Figs. 10.8 - 10.11 all of which are self explanatory with the following short additional notes.

The example concerns the analysis and comparison of distribution system costs and outage costs and the optimisation of the system expansion plan. Six alternative distribution system plans are analysed. Basic plans 1 to 3 correspond to low, medium and high reliability systems. Plans 4 to 6 are hybrids, derived from the basic plans. There are two different versions of plan 2.

The urban area in question experienced economic and demographic growth over two decades, averaging an annual population increase of more than 20% during this period. This growth is likely to be sustained, but at a reduced rate. For the purpose of the study the urban area was divided into geographical cells, within broader geographical areas whose boundaries are represented by the other lines on Fig. 10.9 and 10.10. All initial costs and benefits were in 'efficiency' prices of optimising national economic growth, that is, ignoring income distributional considerations. The 1976 domestic-priced currency was adopted as the 'numeraire' or unit of account because, in monetary terms, more than 99% of the items considered were domestically produced goods and services.

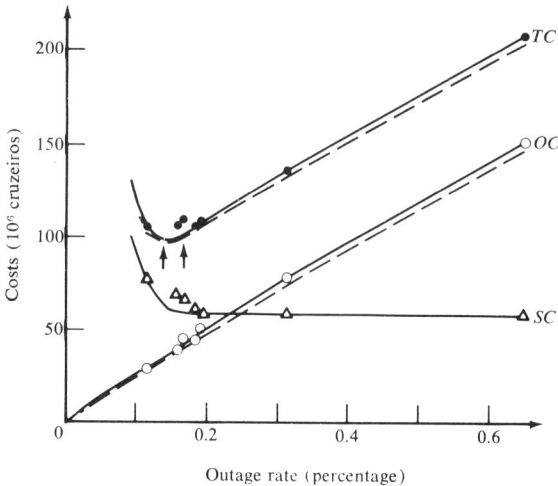

Note: SC = distribution system supply costs; OC = global outage costs; and TC = total costs. The plotted data points and solid lines refer to efficiency priced costs; the broken lines indicate the costs in terms of social prices.

Fig. 10.8   Optimisation of the outage system. Costs versus outage rate

Note: SC = distribution system supply costs; OC = global outage costs; and TC = total costs. The plotted data points and solid lines refer to efficiency priced costs; the broken lines indicate the costs in terms of social prices.

MUNASINGHE, M.: 'The economics of power system reliability and planning'. Fig. 13.1. Published for the World Bank (Johns Hopkins University Press, Baltimore, USA, 1979). Reproduced by permission of the Johns Hopkins University Press

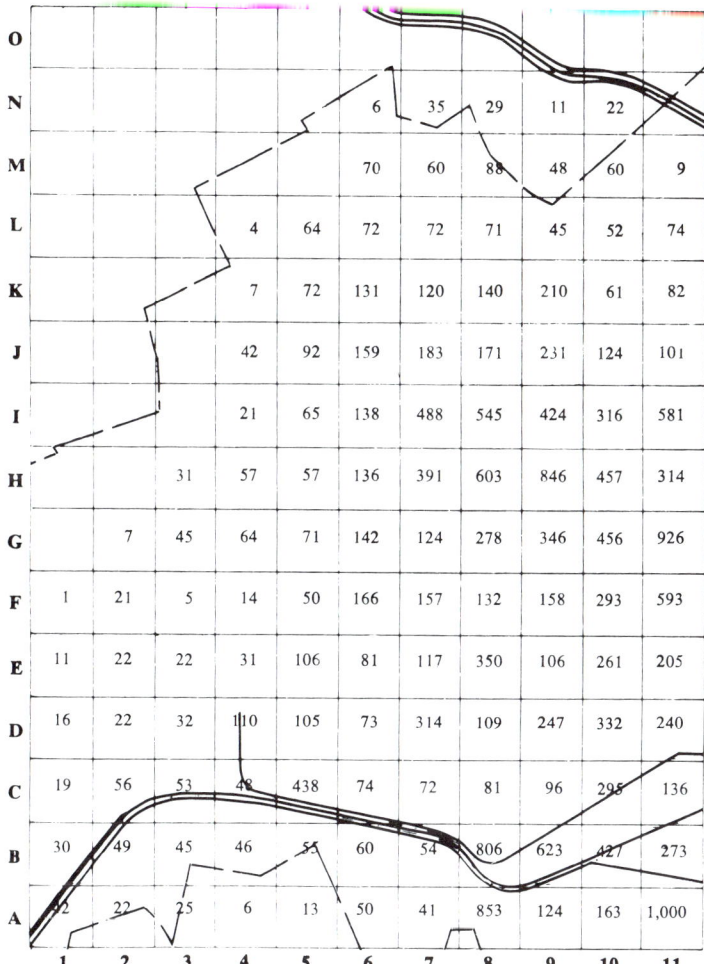

	1	2	3	4	5	6	7	8	9	10	11
O											
N					6	35	29	11	22		
M					70	60	88	48	60	9	
L				4	64	72	72	71	45	52	74
K				7	72	131	120	140	210	61	82
J				42	92	159	183	171	231	124	101
I				21	65	138	488	545	424	316	581
H			31	57	57	136	391	603	846	457	314
G		7	45	64	71	142	124	278	346	456	926
F	1	21	5	14	50	166	157	132	158	293	593
E	11	22	22	31	106	81	117	350	106	261	205
D	16	22	32	110	105	73	314	109	247	332	240
C	19	56	53	48	438	74	72	81	96	295	136
B	30	49	45	46	55	60	54	806	623	427	273
A	32	22	25	6	13	50	41	853	124	163	1,000

Note: Discount rate = 12 percent.

Fig. 10.9   Medium reliability plan B.   Present discounted value of
outage costs by cell (thousands of Cr $)

Note: Discount rate = 12%

MUNASINGHE,   M.:   'The   economics   of   power   system
reliability and planning'.   Fig. 13.2. Published for the
World Bank (Johns Hopkins University Press, Baltimore,
USA, 1979).   Reproduced by permission of the Johns Hopkins
University Press

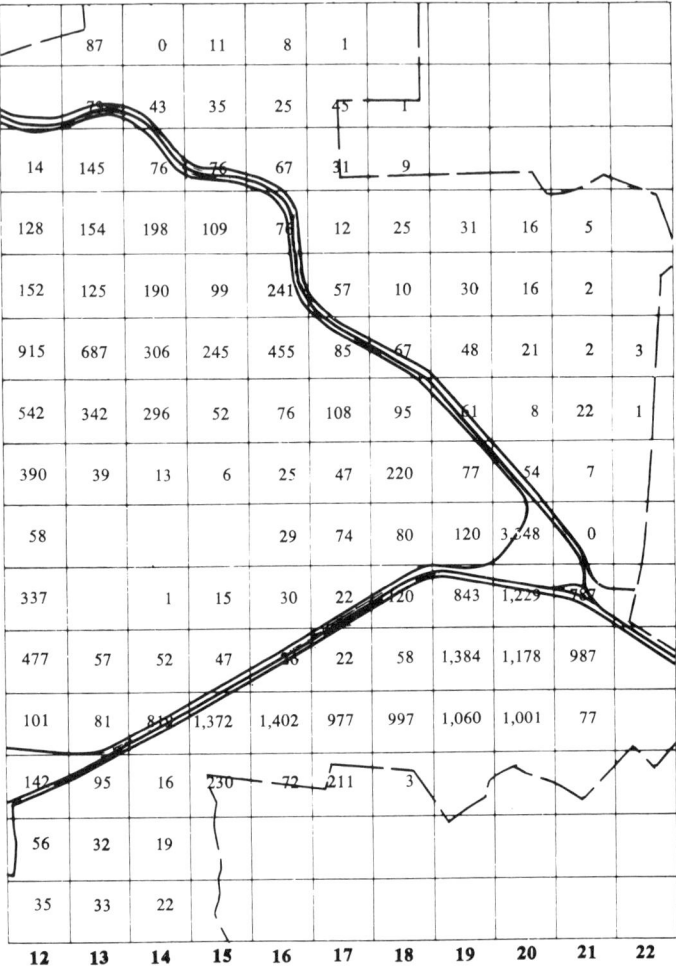

12	13	14	15	16	17	18	19	20	21	22
	87	0	11	8	1					
		43	35	25	45	1				
14	145	76	76	67	31	9				
128	154	198	109	76	12	25	31	16	5	
152	125	190	99	241	57	10	30	16	2	
915	687	306	245	455	85	67	48	21	2	3
542	342	296	52	76	108	95	61	8	22	1
390	39	13	6	25	47	220	77	54	7	
58			29	74	80	120	3,248	0		
337		1	15	30	22	120	843	1,229	987	
477	57	52	47	76	22	58	1,384	1,178	987	
101	81	819	1,372	1,402	977	997	1,060	1,001	77	
142	95	16	230	72	211	3				
56	32	19								
35	33	22								

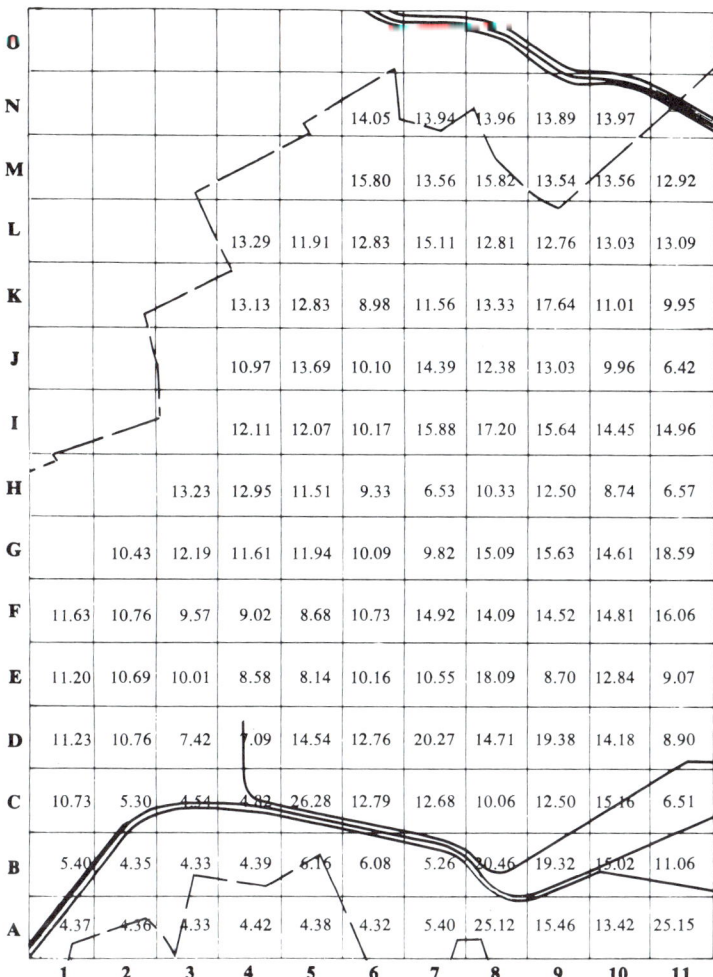

Fig. 10.10 Medium-reliability Plan B. Outage cost per kilowatt-hour lost by cell

MUNASINGHE, M.: 'The economics of power system reliability and planning'. Fig. 13.3. Published for the World Bank (Johns Hopkins University Press, Baltimore, USA, 1979). Reproduced by permission of the Johns Hopkins University Press

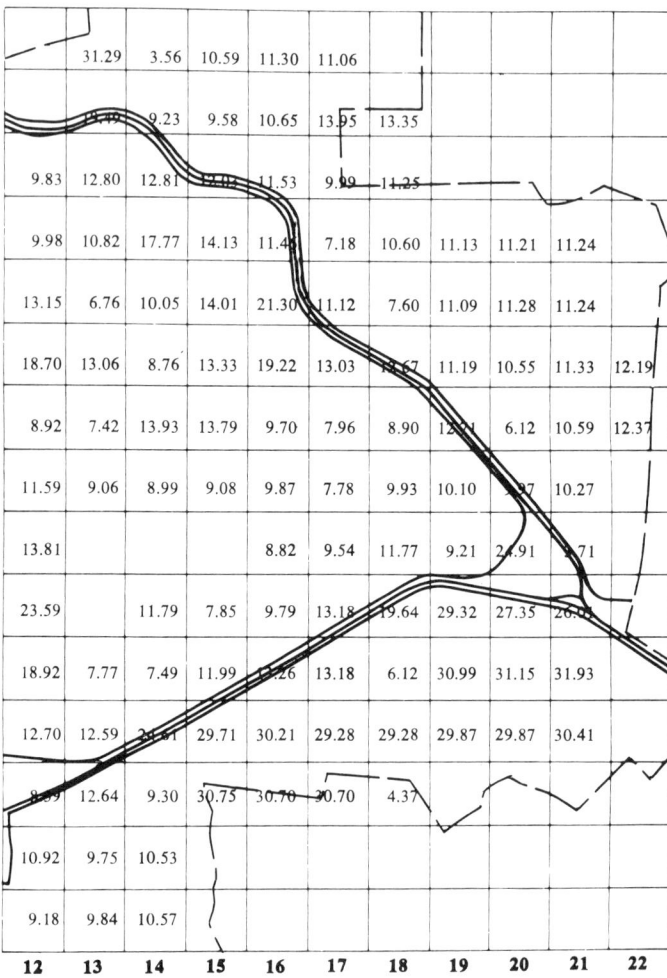

	12	13	14	15	16	17	18	19	20	21	22
	31.29	3.56	10.59	11.30	11.06						
		9.23	9.58	10.65	13.95	13.35					
	9.83	12.80	12.81		11.53	9.99	11.25				
	9.98	10.82	17.77	14.13	11.4	7.18	10.60	11.13	11.21	11.24	
	13.15	6.76	10.05	14.01	21.30	11.12	7.60	11.09	11.28	11.24	
	18.70	13.06	8.76	13.33	19.22	13.03	67	11.19	10.55	11.33	12.19
	8.92	7.42	13.93	13.79	9.70	7.96	8.90	12	6.12	10.59	12.37
	11.59	9.06	8.99	9.08	9.87	7.78	9.93	10.10	7	10.27	
	13.81				8.82	9.54	11.77	9.21	24.91	71	
	23.59		11.79	7.85	9.79	13.18	9.64	29.32	27.35	26	
	18.92	7.77	7.49	11.99	26	13.18	6.12	30.99	31.15	31.93	
	12.70	12.59		29.71	30.21	29.28	29.28	29.87	29.87	30.41	
	9	12.64	9.30	30.75	30.70	30.70	4.37				
	10.92	9.75	10.53								
	9.18	9.84	10.57								

Table 10.1   Global characteristics of the four basic alternative distribution system plans

System Plan	Reliability index (R)	Outage rate, OR)percentage)[a]	Outage costs, OC (10^6 Cr$)[b]	Supply costs, SC (10^6 Cr$)	Total costs, OC + SC (10^6 Cr$)	Outage costs/ kWh lost, OCK (Cr$/kWh)
1 Low-Reliability	0.9935	0.650	150.1	57.5	207.6	14.0
2A Medium-Reliability	0.9969	0.314	77.2	58.6	135.8	14.9
2B Medium-Reliability	0.9981	0.191	49.5	57.8	107.3	15.7
3 High-Reliability	0.9988	0.116	28.4	76.9	105.3	14.8

Note: Present discounted values of quantities from 1976 to 2006 as defined in the text; discount rate = 12%

a   OR = 100.(1−$\underline{R}$) = 100.OE/TE

b   US$ = Cr $ 12.35 (end 1976)

Source: MUNASINGHE, M.: 'The Economics of Power System Reliability and Planning', Table 13.1. Published for the World Bank by The Johns Hopkins University Press, Baltimore, USA, 1979. Reproduced by permission of The Johns Hopkins University Press.

Table 10.2   Energy Use and Outage Costs, by Major Consumer Category

Consumer category	Total energy, TE GWh	Outage Costs, OC (10^6Cr$)[a]				Outage costs/kWh lost, OCK (Cr$/kWh)			
		Plan 1	Plan 2A	Plan 2B	Plan 3	Plan 1	Plan 2A	Plan 2B	Plan 3
Residential	484.1	68.2	34.8	20.5	9.9	17.9	18.2	18.6	17.5
Main industrial	371.8	77.4	40.6	27.8	17.8	31.4	33.5	30.0	26.5
Service industrial	224.7	4.1	1.6	1.0	0.6	3.4	3.2	3.8	3.6
Public lighting	60.4	0.2	0.1	–	–	0.4	0.4	0.4	0.4
Hospitals	11.7	0.3	0.1	–	–	5.9	5.9	5.9	5.9

Note: Present discounted values of quantities from 1976 to 2006 as defined in the text; discount rate = 12%

Source: MUNASINGHE, M.: 'The Economics of Power System Reliability and Planning', Table 13.2. Published for the World Bank by The Johns Hopkins University Press, Baltimore, USA, 1979. Reproduced by permission of The Johns Hopkins University Press.

Table 10.3   Disaggregate energy use and outage costs of residential
and main industrial consumers

Consumer category	Total energy, TE GWh	1	Outage costs, OC (10^{6}Cr$) Plan		
			2A	2B	3
Residential					
Lower	19.9	2.85	1.36	0.83	0.38
Lower-middle	152.3	20.11	10.02	5.98	2.87
Upper-middle	198.9	27.47	14.00	8.27	4.01
Upper	113.1	17.72	9.43	5.46	2.65
Main industrial					
A1	9.7	0.61	0.26	0.14	0.10
A2	30.5	1.11	0.48	0.27	0.19
A3	63.9	9.45	5.29	3.92	2.84
A4	145.2	44.13	22.18	15.54	8.73
A5	88.1	19.40	10.07	5.92	4.40
A6	33.4	26.70	2.28	2.03	1.52

Note: Present discounted values of quantities from 1976 to 2006 as
defined in the text: discount rate = 12 %

Source: MUNASINGHE, M.:   'The economics of power system reliability
and planning', Table 13.3.   Published for the World Bank by The Johns
Hopkins University Press, Baltimore, USA, 1979.   Reproduced by
permission of The Johns Hopkins University Press.

Outage costs/kwWh lost, OCK(Cr$/kWh) Plan				Outage rate, OR (%) Plan			
1	2A	2B	3	1	2A	2B	3
16.5	16.8	17.2	16.2	0.870	0.406	0.243	0.117
16.6	16.8	17.3	16.2	0.798	0.391	0.228	0.117
17.6	17.9	18.3	17.2	0.783	0.394	0.227	0.117
20.6	20.8	21.4	20.0	0.760	0.401	0.226	0.117
24.9	24.9	24.9	25.0	0.269	0.106	0.056	0.041
13.6	12.8	11.9	11.8	0.251	0.124	0.074	0.054
36.7	48.0	73.1	75.5	0.403	0.173	0.084	0.059
31.8	35.2	31.9	25.2	0.948	0.431	0.334	0.237
35.0	32.0	21.9	22.1	0.630	0.357	0.306	0.226
16.6	21.2	22.8	22.8	0.482	0.322	0.267	0.200

Table 10.4   Data for Optimisation of the Distribution System

System plan	Outage rate, OR (%)[a]	Supply costs, SC (106Cr$)[b]	Efficiency prices		Social prices i.e. with social weighting	
			Outage costs,OC (10^6Cr$)	Total costs,TC (10^6Cr$)	Outage costs,OC (10^6Cr$)	Total costs,TC (10^6Cr$)
1 (Low-reliability)	0.650	57.5	150.1	207.6	145.3	202.8
2A (Medium-reliability)	0.314	58.6	77.2	135.8	74.2	132.7
2B (Medium-reliability)	0.191	57.8	49.5	107.3	47.9	105.7
3 (High-reliability)	0.116	76.9	28.4	105.3	27.6	104.5
4 (Hybrid A)	0.184	59.9	44.5	104.4	42.9	102.8
5 (Hybrid B)	0.166	65.2	44.3	109.5	44.1	109.3
6 (Hybrid C)	0.159	67.4	39.3	106.7	39.1	106.5

Note: Present discounted values of quantities from 1976 to 2006 as defined in the text; discount rate = 12 percent

a   OR = $100.(1-R) = 100.OE/TE$

b   US$ = CR$ 12.35   (end 1976)

Source: MUNASINGHE, M.   'The Economics of Power System Reliability and Planning', Table 13.4. Published for the World Bank by The Johns Hopkins University Press, Baltimore, USA, 1979. Reproduced by permission of The Johns Hopkins University Press.

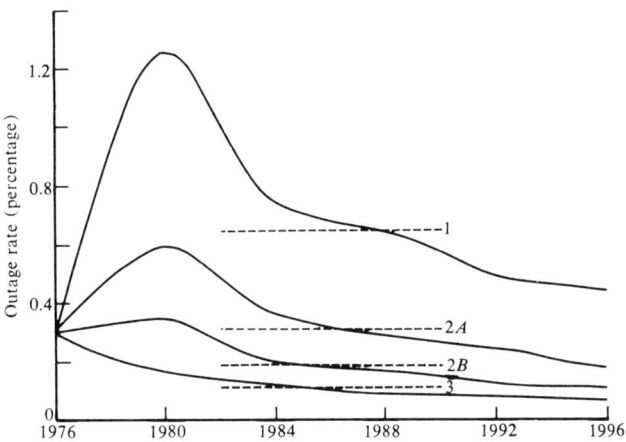

Note: Broken lines indicate the discounted aggregate value of outage rate.

Fig. 10.11   Evolution of outage rate over time

Broken lines indicate discounted aggregate value of outage rate.

MUNASINGHE, M.: 'The economics of power system reliability and planning'. Fig. 13.4. Published for the World Bank (Johns Hopkins University Press, Baltimore, USA, 1979). Reproduced by permission of the Johns Hopkins University Press

Shadow pricing was done in a relatively straightforward way. All domestically produced equipment and materials were valued directly at domestic market price but net of local taxes where applicable. An opportunity cost of capital (OCC) of 12% was selected as the discount rate. The shadow wage rate (SWR) was taken as the same as the market wage rate. This simplified procedure eliminated the need for time-consuming estimation of conversion factors for many domestic inputs to convert the values of these items to 'border' prices, i.e. prices based upon external currencies, most items being produced domestically (see above). Incremental kilowatt (kW) losses in the distribution system were considered as additional capital costs. Hydro-electric generating capacity costs were assumed at US$400 to US$450 per kW, annuitised over a 30- to 35-year lifetime, at a discount rate corresponding to the OCC. Incremental kWh system losses were considered costless, because the system concerned was basically 100% hydro-electric and likely to remain so into the future.
   Fig. 10.8 we have met already in Chapter 3. Fig. 10.11 illustrates the difference in low reliability and high reliability plans,

especially in the early stages of the network development.

10.6 SUMMARY

Networks introduce geography into generation planning together with additional capital and running costs; they also give many more alternative development programmes from which to choose. Choosing between projects/programmes thus becomes more complex, and it helps to consider the various functions which transmission ciruits carry out within a network when testing their economic justification. As transmission voltages get progressively higher, national and inter-national grids become easier to justify technically, financially and economically, and this has an effect on generation planning. Distribution systems tend to grow in a definite least-cost way, starting as radial networks, and progressing through all the stages of interconnection to become fully meshed configurations.   Un-fortunately distribution planning does not deal very well with obsolescence, and it is difficult to justify economically the scrapping of old networks, even though the latter may have gone beyond their physical lives and may be in serious danger of collapse.

REFERENCES

[1]   BERRIE, T.W., DREYFUS, H.B., and KNIGHT, U.G.: 'Primary system
       planning in England and Wales for security of supply'. CIGRE
       1967, paper 21968

[2]   VAN DER TAK, H.G.: 'The economic choice between hydro-electric
       and thermal power developments'. Appendix B, World Bank Staff
       Occasional Paper No. 1 (Johns Hopkins University Press,
       Baltimore, USA, 1974)

[3]   BERRIE, T.W.: 'Prospects for interconnection in south-east
       Asia'. Elect. Rev., 7 Dec. 1979, 205, (22)

[4]   EGGLETON, M.N.: 'Scenarios of network structure'. Paper for
       International Symposium on Electricity Economics and Load
       Management, Imperial College, London, 20-24 March 1980

[5]   CALVERLEY, T.E., WILTSHIRE, T.J., and BERRIE, T.W.: 'The
       development of interconnection between countries of Europe and
       its relevance to SE Asia'. Paper for BEAMA Conference,
       Singapore, 12-15 Nov. 1979

[6]   See Reference 4, p. 3 for details

[7]   Union pour la Coordination de la Production et du Transport de
       l'Electricité (UCPTE) covering Austria, Belgium, West Germany,
       Holland, France, Italy, Luxembourg and Switzerland

[8]   MUNASINGHE, M.: 'The economics of power system reliability and
       planning'. (Johns Hopkins University Press, Baltimore,
       Maryland, USA, 1980)

[9]   Some of the material in this part of the chapter is based on:
       FORD, D.C.: 'Distribution economics'. Paper given at Inter-
       national Symposium on Electric

Chapter 11

Rural electrification

11.1 RURAL ENERGY

11.1.1 Introduction

Rural electrification must be seen within the context of the overall rural energy sector because rising fuel prices are all-important. In the rural areas of most developing countries non-commercial energy predominates; Table 11.1 shows the per capita use of non-commercial energy in rural areas in six developing countries.

In 1975 the total energy requirements of the developing countries were about 18 million barrels a day of oil equivalent (BPDOE). Overall about half of this was met from non-commercial sources, mainly in rural areas. Oil supplied about one third of the commercial energy. In 1979 these total energy requirements were about 23 million BPDOE with a larger percentage of oil.[1] If current trends continue the total amount might be 42 million BPDOE by 1990, with only 60% from commercial sources.[2] For the majority of the 2.5 billion people living in rural areas non-commercial energy sources will continue to play a dominant part in their lives, at least during the next two or three decades. This fact is often overlooked and the following important point forgotten: rural electrification often competes with non-commercial energy and the economics of electrification must be worked out allowing for the environmental benefits of using less non-commercial energy.

11.1.2 Quantity and quality of rural energy

Quality and quantity of energy tends to be low in rural compared to urban areas. This can both improve and worsen the economic case for substituting electricity for other fuels; we must ask ourselves the following type of questions:

(i)   Do rural areas need the same per capita energy as the towns and industrial areas to be progressive?

(ii)   Would not rural areas be content with a lower quality fuel than

Table 11.1 Estimated per capita use of energy in rural areas of six
developing countries in kilojoules per day

Country	India [1]	China Hunan [2]	Tanz- ania [2]	Northern Nigeria [2]	Northern Mexico [2]	Bolivia [2]
Human labour	2.80	2.67	2.67	2.55	3.14	2.97
Animal work	4.18	3.85	-	0.54	5.44	7.66
Fuel-wood	11.97	-	63.09	43.00	40.61	95.58
Crop residues	4.85	57.32	-	-	-	-
Dung	2.80	-	-	-	-	-
Total non-commercial	26.60	63.84	65.76	46.09	49.19	106.21
Coal, oil, gas and electricity	2.22	8.58	-	0.08	82.94	-
Chemical fertilisers	0.92	1.42	-	0.21	22.31	-
Total commercial	3.14	10.00	-	0.29	105.25	-
Total	29.76	73.84	65.76	46.38	154.44	106.21

Source: MASTERS, E.F.O., and BERRIE, T.W.: 'Problem solving using
technology, economics and politics'. World Energy Conferene, Munich,
1980. Reproduced by permission of the World Energy Conference

the towns if the price was right, e.g. fuel-wood instead of kerosene
or electricity for cooking?

(iii) Are not the types of energy use by grade and amount different
in rural areas compared with urban areas?

It is difficult to deal with these points definitively. The
population of countries with large rural areas varies from 0.5 million
to above 700 million; per capita incomes from US$100 per annum to
over US$2,000 per annum.[3]

In most of these countries income distribution is heavily weighted
towards those with low per capita income and often most people live
in rural areas, but this energy picture is not uniform throughout
the world. Certain nations like China and Tanzania have deliberately
implanted industry, sometimes heavy labour-intensive industry, in
rural areas; such a process could fundamentally change the energy
position in general and electricity in particular. Rural electrification would become more economic and the 'special' characteristics
of rural electrification generation, transmission and distribution

Table 11.2 Primary energy sources and electric power: most common applications and possible future applications in rural areas

	Human	Animal	Biomass	Animal waste 2/	Kerosene	Other petroleum fuels
Domestic						
Cooking heat			X	X	X	
Heating			X	X	X	
Refrigeration					X	
Lighting					X	
Radio						
Ironing			X			
Potable water pumping or distillation	X	X				X
Agricultural Production						
Ploughing	X	X				X
Fertilising, sowing	X					X3/
Irrigation	X	X				X
Harvesting	X	X				X
Threshing	X	X	X			X
Drying			X			X
Grinding, pressing	X	X	X			X
Transport	X	X				X
Storage						
Input to energy conversion process			X1/	X		X

1/ Anaerobic production of methane gas and generation of steam or electric power

2/ Excluding use as a fertiliser

3/ This includes the use of petroleum fuels as feedstocks for the manufacture of fertilisers

Source: MUNASINGHE, M., and WARREN, C.J.: 'Rural electrification, energy economics and national policy in the developung countries'. World Bank Paper, Washington DC, USA, 1979. Reproduced by permission of the World Bank

Table 11.2   Primary energy sources and electric power: most common applications and possible future applications in rural areas (continued)

	Coal	Liquid gas Biogas	Wind	Water power	Solar	Grid or generator set electric power	Batteries
Domestic							
Cooking heat	X	X			X4/	X	
Heating	X	X			X	X	
Refrigeration		X			X5/	X	
Lighting		X				X	
Radio						X	X
Ironing	X					X	
Potable water pumping or distillation			X	X	X6/	X	
Agricultural Production							
Ploughing							
Fertilising, sowing							
Irrigation	X		X		X7/	X	
Harvesting							
Threshing				X		X	
Drying	X				X8/	X	
Grinding, pressing	X		X	X		X	
Transport							
Storage							
Input to energy conversion process	X		X	X	X		

4/   Solar cookers can probably be discarded as a social failure as people do not like cooking in this unfamiliar way

5/   Present high cost is a deterrent to individual household use. More applicable to community scale systems

6/   Solar distillation has high capital costs, but for large quantities can be cheaper than fuel-based methods, depending on the climate

7/   Applications of solar energy to irrigation have not been very successful as yet. Photovoltaic cell-driven-pumps are expensive

8/   The traditional method is to dry the crop by spreading it on the ground. This can lead to high crop losses. Improved methods of using solar energy to dry the crops while in the storage bins are being developed.

(see later) would be restored to 'normal'.

Table 11.2 based on Reference[4] shows the most common applications of energy by end usage in rural areas. Table 11.3 shows Table 11.1 in another form, to include energy usage. All three Tables are important. For subsistence farming, e.g in those parts of Bolivia, Nigeria and Tanzania where non-commercial energy provides almost all the energy required, the only commercial energy demanded is oil and then only for cooking; there is no place for the high-grade, expensive form of energy called electricity. However, in parts of Mexico where intensive agriculture using high-yield grains is paramount, there is an increasing demand for energy of a much higher grade and particularly for electricity. There are many intermediate cases, especially in India and China, which together constitute over two-thirds of the total world population and over three-quarters of the world rural population.

Table 11.3   Estimated per capita use of energy in rural areas of six developing countries (including end usage) in kilocalories per day†

Country	India	China, Hunan	Tanz-ania	Northern Nigeria	Northern Mexico	Bolivia
	[1]	[2]	[2]	[2]	[2]	[2]
Human labour	0.69	0.65	0.65	0.62	0.76	0.72
Animal work	1.00	0.93	-	0.14	1.32	1.84
Fuel-wood	2.88	-	15.09	10.28	9.72	22.85
Crop residues	1.16	13.70	-	-	-	-
Dung	0.69	-	-	-	-	-
Total non-commercial	6.42	15.28	15.74	11.04	11.80	25.41
Coal, oil, gas and electricity	0.54	2.06	-	0.03	19.82	-
Chemical fertilisers	0.22	0.35	-	0.06	5.34	-
Total commercial	0.76	2.41	-	0.09	25.16	-
Total	7.18	17.69	15.74	11.08	36.89	25.41

† Source: Derived by the author from Table 1 and the following: [1] REVELLE, R.: Chapter on 'Requirements for energy in the rural areas of developing countries' in BROWN, N.L.(Ed.): 'Renewable energy resources and rural applications in the developing world'. American Assoc. for the Advancement of Science, Washington DC, 1977; [2] MAKHIJARI, and POOLE: 'Energy and agriculture in the Third World' (Ballinger Publishing Co., Cambridge, Mass., 1975)

11.1.3 Pattern of rural energy use

The pattern of energy use in rural areas is quite different from that in urban areas mainly because life in rural areas is intensely agricultural, which is unfortunate for financial and economic viability of all village energy projects, since it leads to low utilisation of most energy production equipment; as low as 10%, particularly where cooking is done by non-commercial fuels.

Measured, therefore, in terms of willingness-to-pay, economic benefits might be small for a very large capital outlay in rural energy projects, e.g. private diesel-electric generators, private diesel motors, grain driers, etc., as has been the experience in India, Pakistan, Bangladesh and other countries.[5] However, patterns of rural energy use and the rural ways of life are beginning to change, as communication improves and the world becomes smaller, towards the patterns of urban areas. This is a long-established trend in developed countries and it is becoming apparent in developing countries, where the rural population is demanding and starting to receive larger quantities and higher grades of energy than one would have considered possible even twenty years ago.[6] In the Pearson Report[7] in 1967 reference is made to government apathy in most developing countries to rural areas even as late as 1965; since 1973 there has been increased interest in rural energy, mainly because of the increasing price of fuels and the growing consciousness of the environmental costs of using non-commercial fuels.

In 1977 the annual per capita subsistence energy requirement in most rural areas was estimated to be about 5×10^{-3} BPDOE, i.e. between 3% and 5% of the average in industrial areas. A tripling of this perceived minimum standard is expected by about 1983.[8] The rapid increase in the quality of life which goes with higher expectations makes the case for improved energy projects in the 1980s much more plausible than in the 1960s and 1970s, especially in 'green-field' situations where there is little commercial energy yet being used.

The following factors will force the use of higher grade energy in rural areas:

(i)   the use of firewood causes deforestation with its disastrous effects

(ii)   the use of dung deprives the land of much needed fertiliser, the artificial forms of which are extremely expensive

(iii) lastly, up to two-thirds of a working day can be spent gathering non-commercial fuels which leaves less time for other work.

11.2 RURAL ELECTRIFICATION

11.2.1 Introduction

Large rural electrification programmes were launched in the developed countries immediately following World War II. In the UK the newly created Electricity Boards were charged by statute to carry out rural electrification, although it was not clear how these were to be financed, e.g. by cross-subsidisation from other consumers or high connection charges for rural consumers.

In the USA the government authority for rural electrification helped to electrify all but the most isolated farmsteads by providing technical assistance and special financing facilities for groups of farms and farming co-operatives. Similarly programmes of rural electrification were instituted by the state in France, this time with some direct government subsidies. There are many other examples.

No attempts were made to economically justify these programmes in the manner now expected by government departments in developing countries and the international lending agencies (see later).

In the developed world almost all potential rural electricity consumers are now connected; in the poorest developed countries work is proceeding rapidly as a matter of governmental policy to connect up all but the most inaccessible hamlets. Once more, no explicit economic justification is normally being called for.

However, in developing countries, especially where capital is short, an economic justification of a rural electrification programme is called for by the government's own economic planning agency and by the international lending agencies providing the funds, e.g. the World Bank, because:

(i)   The rules of the lending agencies stipulate investing only in economically justified projects.

(ii)   Rural electrification requires a great deal of capital investment per consumer.

(iii)   The low revenues from rural electrification reduce the possibility of economic 'take-off'.

11.2.2 Special characteristics of rural electrification

We now show the special characteristics of rural electrification and their effects on the power utility.

RURAL ELECTRIFICATION CHARACTERISTICS

(i)   Low proportion of potential consumers initially connected, typically 20% to 30%

(ii)   Low growth rate in the number of new consumers connected, typically 2% to 3% per annum, at least in the early years

(iii)   Low initial use of electricity per consumer, typically 200 kWh to 300 kWh per annum, at least in the early years

(iv)   Low initial growth in electricity consumption, typically 3% to 5% per annum

(v)   Low revenue from consumers, at least in the early years

(vi)   High capital cost of installation per consumer for the utility and possibly the consumer, typically up to ten times that of urban consumers

(vii)   Low utilisation of equipment belonging to the power utility

(viii) Low load factor of the consumer equipment, typically 30%

(ix)   Often a low power factor, because of the frequent dominance of a single motor drive in a village

(x)   High operating, maintenance and fuel costs per kWh sold

(xi)   Poor voltage and quality of supply

(xii)   Disinclination for changing over to electricity from substitutes, e.g. kerosene, because new equipment needs to be purchased and kerosene sales are often subsidised more than electricity sales, especially in rural areas

(xiii) Poor arrangements for finance offered to householders and agro-industry to pay for equipment, connection charges, housewiring, etc.

(xiv)   Difficulties of collecting revenues

These basic characteristics make difficult demands on the utility
and the government because:

(i)   The low annual financial return on turnover means that the
        utility, the government, or both, must subsidise rural
        electrification for a good number of years; possibly for ever

(ii)   The low long-run internal financial return on investment makes
        the project difficult to justify to the utility or to the
        government

(iii)   The economic return to the economy is difficult to quantify
        because some cost data is completely absent, or unreliable,
        and the economic benefits are difficult to measure. Often
        the economic return is as low as the long-term internal
        financial return

(iv)   There can be institutional problems because of the practical
        difficulty in setting the subsidy levels and because of the
        many different authorities involved in every aspect of the
        work

(v)   The pricing policy set for the rural areas might have to be
        the same as for all areas of the country regardless of
        imbalances in costs; e.g. because a uniform tariff policy
        legally applies throughout the country.

11.2.3 Ground rules for rural electrification

Although rural electrification, like rural water supply, is not
necessarily meant to be viable financially either to the government
or to the utility concerned, they must always be proven to be: (i)
economic with respect to the community,[9] and (ii) the most cost-
effective way of carrying out the service. At least, this is the
view of the lending agencies who lay down ground rules for appraising
rural electrification projects.[10]

Cumulative investment in rural electrification in the World Bank's
areas of operation were about US$10 billion in 1971, i.e. about 10%
of the total investment in electric power, see Reference[10]. In the
last ten years the pace has accelerated; about US$20 billion has been
spent or committed, and about 25%, or about 350 million of the people,
now receive electricity in rural areas.[11] The percentage for urban
areas is about 75%.

The limited post-evaluation work of the late 1960s and early 1970s
showed that economic benefits were poor, poorer than some had
originally forecast;[12] rural communities were not taking off
economically, even under the best circumstances. Researchers in the
late 1970s believed that this was more likely to be caused by other
infrastructures failing to develop rather than failure of the rural

electrification sector. For example, many rural development
specialists blamed poor communications for the failure of many
schemes. We now give some basic ground rules for rural electrification projects based upon World Bank publications.

BASIC GROUND RULES FOR RURAL ELECTRIFICATION PROJECTS

(i)   Connect only those consumers whose benefit to the economy of
being connected exceeds the cost to the economy. This is often
a difficult calculation but the World Bank guidelines[13] have
been successfully applied by planners.

(ii)   Use a quality of supply which is appropriate to the circumstances of:
(a) being safe, but
(b) saving as much capital investment and especially foreign
exchange as possible, thereby
(c) ensuring that the maximum number of consumers meet the
economic tests and are in fact connected; which means
(d) using a quality of supply of both apparatus and systems
below, and possibly well below, that used in urban areas.

(iii) Encourage the substitution of electricity for other fuels by
end-usage where this is beneficial to the economy. For
example, encourage the substitution of electric lighting for
kerosene lighting because this usually makes a resource cost
saving to the economy, i.e. saves on the nation's fuel bill.
Consumers often pay more than the electricity tariff for
substitutes, which indicates the minimum level of the
willingness-to-pay or 'consumer surplus' of a consumer, i.e.
a measure of the benefit to the consumer of having electricity
over and above that indicated by the electricity tariff. This
enables a truer measure of the benefits of electricity to be
gained than indicated by the potential revenues; [(i) above
refers].

(iv)   Carry out as extensive a survey as possible of the rural areas
concerned. Devising and analysing questionnaires needs skill
and care. Data are needed to find out if the proposed scheme
will 'take-off' economically within the foreseeable future.
Such data will concern population level and growth; income
level and growth; occupations; present means of providing
essential services like lighting, cooking, heating, cooling,
water supply; food preservation; communications; education;
medical facilities etc. The data are used for a rough
macro-economic analysis to compare the area with other similar
rural areas which have taken-off economically.

(v)   Use indigenous fuels if these are at all likely to be economic
to the economy, especially those fuels indigenous to the rural
areas in question. Thus micro-hydro-electric schemes should
be encouraged, whereas the use of fuelwood, dung and vegetable
wastes for cooking should be discouraged.

11.3   APPENDIX A: RURAL ELECTRIFICATION STUDY. WORK PROGRAMME FOR ECONOMICS[14]

11.3.1   Establish the marginal costs of supply

Sufficient information will probably have been assembled from initial studies of forecasts of maximum demand (kW) and electrical energy (kWh) to enable a preliminary network structure to be postulated. From this network structure and the likely generation-programme preliminary marginal costs of supply should be worked out per kW and per kWh for supplying new loads. These marginal costs of supply will be related to geography, size of load (by group), type of load (by class), load operating regime etc. up to (say) 25 years hence, the end of the forecasting period.

The preliminary network structure will be modified by successive approximation as loads are shown to be economic, or otherwise, to supply. Not more than one or two iterations should prove necessary except for special sections of the network.

The marginal costs of supply should be kept as simple in format as possible to within the limits of accuracy of the method used and of the data on which they are based. They will be used to calculate economic cost-to-benefit ratios for determining whether, on economic grounds, each major load, or block of load by type, should be connected to the public supply. The marginal costs of supply will thus be in economic rather than in financial terms. Thus load forecasts should be built up of demands for electricity which are controlled by the economics of cost-benefit analysis. At this stage we do not know whether interconnection between loads or groups of loads is the most economic solution.

These load forecasts will supplant any earlier ones and be used for recommending development programmes.

11.3.2   Economics of connecting pumping loads, flour mills, motor drives etc.

There are several technological choices for water pumping (tube-wells), for flour milling and motor drives. Choosing the most economic option should be done by determining the least-cost solution between:

(i)   using diesel engines to drive the pumps, flour mills, or other drives
(ii)   using small local diesel generators to supply electric motors (on the surface in the case of the pumps) to drive the pumps, or the flour mills, or motor drives
(iii)   the same as (ii) but using submersible pumps
(iv)   using the mains supply to feed electric motors (on the surface in the case of the pumps) to drive the pumps, or the flour mills, or motor drives
(v)   the same as (iv) but using submersible pumps.

If the least-cost solution is to use the mains supply, then an economic cost-to-benefit ratio calculation should be carried out for each major pumping load, flour mill or motor drive, or group of these, to determine whether that particular load should become part of the load forecast (see earlier).

11.3.3   Economics of auto-generation

In a rural situation there are often several in-house generating sets, mainly the property of hotels, small industries and local villages.   The economics of supplying existing and possible future loads in this category should be determined, to examine whether these loads should also be added to the load forecast rather than left on private supply.

There will not normally be many technological options; so the choice between technologies can be made at the same time as calculating whether the loads should be connected.

The criterion for deciding whether the loads should be supplied by captive plant or the mains will be the economic cost-to-benefit ratio.

11.3.4   Quality of supply

The quality of supply appropriate to a particular load will vary in accordance with the importance of the load to the economy.   The appropriate standard will vary between one which is 'adequate', for non-critical loads such as domestic, and one which is 'high', for pumping and agro-industry.

The quality of supply appropriate for each class and type of load should be determined by the reliability rule (Chapter 3), i.e. an examination of incremental economic outage costs and incremental economic system costs as the quality of supply is increased or decreased from a 'norm'.

11.3.5   Any expected development of tourism?

Tourism development and its associated load is a major factor in the development of many rural economies today.   This item is therefore important and should be researched especially for the following:

(i)   likely airport facilities and international/national flight patterns

(ii)   likely infrastructure development: services

(iii) likely superstructure development: hotels

(iv)   likely port facilities

(v)   likely kind of tourist attracted: cross-section by income,
      ethnic group, interest, age, point of origin, length of stay,
      etc.

(vi)   financial incentives offered by government to developers

(vii) economics of foreign-exchange inflows and outflows due to the
       development of the tourist industry

(viii) intrinsic benefits of tourist industry: income distribution,
       secondary benefits, multiplier effects, etc.

11.3.6   Resource capabilities

The likely rate of rural development will mainly depend on the
resource capabilities of the country. These capabilities are in two
major categories, physical and financial:

(i)   Physical resources will be those made available by government
      authorities, public utilities and by private authorities.
      These include materials, labour, management, transport, con-
      struction services, accommodation, back-up expertise, e.g.
      critical-path analysis, etc.

(ii) Financial resources will be those made available by government
      authorities, public utilities and by private corporations,
      e.g. hotel owners and other private investors and developers.

11.3.7   Calculation of economic benefits

Financial benefits of connecting new consumers are measured by the
revenues collected plus connection charges and any consumer contributions. Such revenues depend on the existing tariff and are
probably a minimum measure of the economic benefit of connecting a
particular demand for electricity to public supply. If tariff levels
are too low or their structures do not reflect long-run marginal costs
of supply, consumers may be willing to pay more (see below).

Besides the financial benefits there are additional benefits to
the economy of connecting new consumers and the accuracy of these
benefits should only be determined as far as the data can be relied
upon.

The most important economic benefit, in addition to financial, will
be any 'surplus' value to consumers over and above the value
represented by their presently proven willingness-to-pay through the
existing tariffs. To calculate this, electricity sales against

cost of production or price 'demand' curves should be drawn for the major classes of consumer, using:

(i)   cost of production of small isolated rural and larger captive plants for industries and hotels

(ii)   cost of production of electricity alternatives, e.g. batteries, accumulators

(iii) price being paid for substitutes of electricity for lighting, heating, cooking by commercial fuels - kerosene, gasoline, acetylene, butane, fuel oil etc.

(iv)   economic and conservation costs for using non-commercial fuels as substitutes: fuel-wood, charcoal, dung, vegetable wastes.

Multiplier effects should also be determined as far as this is possible, e.g. the increased financial and physical tempo of economic development due to electric supply.

11.3.8 Existing and future likely tariff levels and structures

Existing tariff levels may be too low in respect to:

(i)   comparison with similar countries

(ii)   whether tariffs date from a time when oil was much cheaper or if the energy policy has substantially changed since then

(iii) surveys of lending agencies like the World Bank seriously recommending tariff increases.

Tariff structures should be based upon marginal costs. The cost of supply information, mentioned above, for each new load should enable this problem to be adequately addressed.

It is important to deal sufficiently with existing and future likely tariff levels and structures to decide whether:

(i)   Existing tariffs indicate less than total benefits ascribable to electricity.

(ii)   Existing willingness-to-pay reflects a low measure of true benefits.

(iii) Likely future increases (decreases) in tariffs would decrease (increase) the amount of electrical energy sold and change the load forecast.

(iv)   There are built-in subsidies to the supply industry which have major economic impact and might affect how to value economic costs and benefits.

(v)   Secondary benefits will be taken care of in any tariff increase
        likely.

11.3.9 Analysis of consumer survey

The first stage is an extensive survey of consumers and potential
consumers. Any work will be materially affected by the analysis of
the survey to indicate present and likely:

(i)   willingness-to-pay and ability-to-pay for electrical energy
        and substitutes

(ii)   direct and indirect benefits of electricity

(iii)   propensity towards obtaining and utilising electricity

(iv)   income distribution elements

(v)   consumption patterns and values placed upon consumption

(vi)   value-added due to electricity

(vii)   cost of loss of connected supplies in economic terms

(viii) financial institutional arrangements to help with connection
        charges, cost of house wiring, installation charges, by
        public and private development or by commercial banks,
        co-operatives, consumers' associations

(ix)   social values of electricity, education, culture, literacy,
        health, hygiene, law-and-order, communication,
        community spirit, breakdown of social barriers, etc.

The results from the survey should be presented in detail at the
end of the study. Meanwhile, results obtained during the analysis
should be presented as evidence to back up any recommendations made
during the course of the study.
    An important item it is hoped to discover from the consumer survey
is the degree of penetration of electricity into existing types and
groups of load: industry, commerce, domestic, agriculture, cities,
towns, villages, farm groups; also from the analysis of the consumer
survey to determine a likely penetration rate for these and new types
and groups of load.
    In many cases the rate of growth in level of demand for connected
consumers may be negated by a high rate of new connections since new
consumers usually have a very low demand to start with. This applies
to new industry, new commerce, but especially to low-income domestic

areas of the villages.

The present and likely rate of penetration is a good indicator of whether making the service of electricity available to a group or type of consumer will enable that service to 'take-off' in an economic sense.

11.3.10   Independent sales and demand forecasts

Initially, load forecasts will normally be made using only one methodology, i.e. by adding together the estimated sales of electricity (kWh) for each consumer class and then using a general load factor to forecast the system simultaneous maximum demand (kW). Some backtracking will have been made to make sure that there was correlation between the assumptions made in forecasting for each consumer class.

Ultimately, it is necessary to make forecasts of sales (kWh) and maximum demand (kW) by more than one method; for cross-checking purposes, load forecasts play such a vital role in rural electrification planning by:

(i)   Making independent forecasts of maximum demand (kW), i.e. not assuming a load factor for converting sales to maximum demand. This should be done by working forwards from the average installed kW capacity of apparatus for each consumer class, using diversity factors between

(a) apparatus in each home or premises
(b) consumers in each class
(c) different consumer class
(d) different geographical areas.

(ii)   Trending of consumer numbers by class, area, consumption (kWh), installed capacity (kW), maximum demand (kW) all by per consumer (or groups of consumer).

(iii) Correlating past sales and maximum demands per consumer class, per area, with macro-economic trends; and using these correlations to forecast sales and maximum demands from development postulated in 5-year plans, tourism development plans, agricultural development plans, etc.

11.3.11   Long-range load forecasts

The result will be a set of load forecasts, low, medium and high, for all areas and separate systems, of sales (kWh) and maximum demand (kW) for each year up to and including, ten years ahead and then for every fifth year up to and including twenty-five years ahead.

These load forecasts should then be compared with the actual and forecast loads for similar areas in the same state of development in the same and other countries.

11.3.12   Methodology to be used for economic justification

The precise methodology to be used should be formulated from such items as:

(i)   What 'numeraire' or common unit of cost to use, usually the net increase in national welfare, the economic efficiency criterion as used in this book.

(ii)   Which conversion factors to use and how to work them out, e.g. to deal with the differences between real and money terms.

(iii) How to take into account budgetary, financial and physical resource constraints.

(iv)   Whether to use investment 'premiums' or put 'premiums' on the benefits from 'productive' users.

(v)   Whether to work in net present-value terms at the 'efficiency' discount rate, or in terms of required rates of return, or both. Whether to work in cost-benefit ratios, or first year returns, or pay-back periods, or present-value pay-back periods, etc.

Economic choices between different technologies to carry out exactly the same function should be worked out using incremental capital costs and total system running (operation, maintenance and fuel) costs of the alternatives as in Chapter 8. For choosing the least-cost solution, the benefits to the economy of investing in any of the alternatives 'per se' should not be taken into account; but the full economic benefits of the least-cost technology must be taken into account when evaluating its net present value or its economic return, as in Chapter 8. 'Shadow-pricing' should be used for both finding the least-cost solution and the economic return on the least-cost solution.

Sensitivity testing on the conclusions reached should be carried out with respect to variations in the main parameters, e.g. fuel costs (as for Section 11.3.13), investment costs (foreign exchange and local), labour costs, foreign exchange rate, some conversion factors affected by social changes (income distribution weightings, premiums on production, emphasis on development).

11.3.13   Fuel costs and energy policy

The price of crude oil should be taken at the border price, i.e. the world market price after making allowance for transport and handling. The prices of oil products should be taken at values derived from the assumed price of crude oil. As there is yet no world market price for coal, a surrogate for this must be taken.

Close surveillance of the level of the world price of oil should be made; also how the assumptions concerning the prices of primary fuel (for power stations mainly) and for secondary fuel (for substitutes for electricity) fit into the declared country energy development policy, e.g. for hydro and geothermal plant, and the energy pricing policy of the government.

From the above the fuel prices to be used will develop logically.

11.3.14 Justifying particular projects in the development programme

Bringing together the 'demand' side with the cost of 'supply' side, with possible iterations with respect to the load forecasts mentioned above, will give recommended phased stages of development for rural electrification up to and including the year (say) 1990, with general indications for the next five years ahead and possibly even for the next ten years ahead; design year 1981.

Particular projects which will be given very close scrutiny, and either justified or rejected, will be:

(i)   further small diesel generating units at existing power stations

(ii)   small or medium hydro-electric stations

(iii)   possible non-conventional and geothermal generating stations

(iv)   connection of the rural development area with elsewhere

(v)   development of transmission systems

(vi)   development of distribution systems.

11.3.15 Capital budgets

In the early stages provisional capital budgets for the next three financial years must be given.

In subsequent work these capital budgets will be 'firmed up' and budgets for all financial years up to and including 10 years ahead should be made.

11.4 APPENDIX B: ECONOMIC RETURN OF A RURAL ELECTRIFICATION PROJECT: A WORKED EXAMPLE[14]

11.4.1 Introduction

It is assumed that the costs* of electrification of the rural

---

* Costs are defined as (i) the capital and other fixed costs of the new investment plus (ii) the operating, maintenance and fuel costs of the total system. Benefits are defined as the worth to the economy

electrification scheme have been fairly well defined during the
technical evaluation of the scheme and the determination of the
least-cost means of meeting the electricity demands. The benefits
of the scheme are much more difficult to determine but these are
required for calculating an economic return on the scheme. At present
there are small isolated electricity utilities in the area, and the
tariff charged by these concerns is the only readily first available
measure of benefit, in the form of willingness-to-pay for
electricity. Current tariffs are of the order of 50 (monetary)
units/kWh for lighting and fans and 25 units/kWh for motive power,
together with a fixed charge or minimum charge which can be expected
to raise the total average charges per kWh a little above these
figures. Relatively few consumers presently use electricity for
anything other than lights, fans etc., so that the above tariffs give
little indication of the full willingness-to-pay for irrigation
pumping loads etc. The existing tariffs, however, are a guide to
benefits for domestic and commercial consumers.

11.4.2 Domestic and commercial consumers

For electricity consumed by domestic and commercial consumers, the
figure of 50 units/kWh may be taken as the basis for the lowest measure
of the benefits of electrification. There are a number of arguments
for using either a higher or a lower figure, and these are summarised
below:

(i)   Some arguments for a valuation above 50 units/kWh

   (a) The standard of supply provided under the scheme for
       electrification could be greatly improved over that for
       which users are currently paying at least 50 units/kWh.
   (b) In at least one of the undertakings, a minimum monthly
       charge of 300 units for shops and 500 units for domestic
       consumers is charged. Together with fixed charges (meter
       rents, minimum charges etc.) true average price levels are
       in fact above 50 units/kWh.
   (c) At present tariff levels there are waiting lists of
       consumers seeking connection.
   (d) Incomes have risen substantially with earnings repatriated
       from other parts of the country and from abroad, and this
       trend is expected to continue with the rise in agricultural
       output foreseen.

---

of the output from the scheme. A minimum value of these benefits
is the revenues directly attributable to the scheme. On the other
hand, consumers may be willing to pay more, and there may be secondary
benefits.

(ii)   Some arguments for a valuation below 50 units/kWh

    (a) After the full electrification scheme, a much larger number
of customers will be connected. For those outside the main
towns, no proof has been made available of willingness-
to-pay (although there is usually some small scale private
generation).

    (b) At present, 25 units/kWh is charged for refrigerators;
presumably some of them at least are likely to come under
the domestic and commercial tariff.

    (c) Tariffs in other parts of the country are considerably
lower than 50 units/kWh, and there may be consumer
resistance to such a rural tariff on grounds of fairness.

On balance, it seems that an average for domestic and commercial
consumers of 50 units/kWh is, however, a reasonable measure of what
consumers under the scheme would be able and willing to pay. However,
there are a number of unquantifiable reasons why this measure is
likely to understate the full economic benefit of electricity to these
consumers.

Firstly, the simple addition of expected revenues makes no
allowance for the 'consumer surplus' representing the net benefit
of electricity to the consumer over and above what is actually paid.
For example, those consumers who presently generate their own
electricity and who find it economic to switch to public supplies
will, in fact, benefit by the amount proportional to what it presently
costs them to generate, a higher value than the price charged by public
supply. Also, electrification of the rural area will provide greater
incentives for young people to remain in the area by improving the
quality of life. Emigration of young, skilled workers is a problem
in rural areas and electrification will at least help to make the
area more attractive.

The rural areas are basically agricultural with average incomes
which, though rising, are below those of the urbanised areas of the
country. Electrification may, therefore, be considered desirable
from the viewpoint of distribution of income between the agricultural
and urban areas of the country. Electrification can be expected to
stimulate further development in the rural areas, both for agro-
industries and cottage industries of all kinds.   It will also
stimulate social and institutional development.   For example,
electric lighting allows people to work longer hours and allows other
activities in the period after dark. The scope for improvement of
health in the area will also be increased by electrification of health
centres and clinics. Electrification provides a more efficient
conversion of chemical energy into heat by, for instance, replacing
paraffin stoves for cooking. In addition, the range is extended.
All these unquantifiables should be taken into account in a proper
evaluation of the benefits of the electrification scheme.

A number of developments are taking place in the rural areas, including the construction of roads, agricultural developments and new agro-industrial projects. There has been an influx of money into the area due to earnings from other parts of the country and abroad, and improved agricultural yields expected in the future. Incomes are rising and should continue to rise in the foreseeable future. The area fulfils many of the criteria required for rural electrification, and there seems little doubt of the very real benefits which will accrue to domestic and commercial consumers if the scheme proceeds.

11.4.3 Irrigation and industrial consumers

There are at present very few irrigation pumps powered by electricity in the rural area and these are charged a rate of 25 units/kWh. In fact, pumping in most of the existing shallow wells uses small diesel engines to provide power. This situation, however, is probably due to deliberate policies and to the siting of existing equipment, rather than because of a choice between the use of electricity for pumping at the above tariff levels and use of diesels. The possible arguments which may be presented to determine the value of electrical energy supplied for irrigation pumping are:

(i)   The long-run marginal economic cost may be used as a first measure of the benefit.

(ii)   The tariff derived from historic costs in financial terms may be so used.

(iii) The cost of extracting water from the present system of shallow wells can be used as a proven measure of willingness-to-pay for extracting water 'per se'.

(iv)   The next-best alternative method of extracting water from the deep tubewells can be used for comparison. The value of electricity to the consumer should be below the alternative cost of diesel-driven pumps, or else the rational consumer would prefer the diesel pumps.

Of these four alternatives, the first two are not really appropriate. Existing tariffs or costs could be used as a measure of benefits, but since such a kind of tariff does not presently exist, the estimated tariffs are not necessarily a good guide. Of the other two alternatives, the value placed on electricity under (iii) would certainly be considerably higher than under (iv). At the present time consumers would be willing to pay up to the full costs presently incurred in extracting water, so this should be a valid measure of the benefits of electrical pumping. But drilling the tubewells will change the situation. With a deep well, the costs of diesel pumping would be substantially reduced, and it is arguable that the costs of diesel drive pumping of the deep tubewells set an upper limit on the value of electricity used for pumping.

The total industrial load was estimated to be some 600 kW in 1980 and 1500 kW in 1981. These figures compare with only about 300 kW at the time the sums were done (1978). Certainly it seemed unlikely that some of the larger industries involved could expect to obtain electricity from existing concerns without a rise in the present tariff to cover the costs of additional equipment. One processing factory for example would have to install its own generation equipment, so that the value of electricity to it can be expected to be well above the present power tariff of 25 units/kWh. Perhaps 50 units/kWh might be appropriate as a measure of the value of electricity for this particular industry. On balance, an average value of around 35 units/kWh might seem a reasonable figure to put on the benefits of electricity used for industrial purposes.

11.4.4 Sensitivity of rate of return to valuation of benefits

The calculations are shown in Tables 11.4 - 11.7. The economic rate of return on the electrification scheme is very sensitive to the evaluation put on the benefits. Table 11.6 shows just how sensitive the calculated economic return is to the valuation put on electricity used for irrigation and industrial purposes.

The diesel-drive alternative to electricity (Table 11.7) was costed to provide a guide to the proper valuation of electricity used for pumping. This valuation uses economic costs, i.e. world prices for skilled labour and fuel and is wholly in real terms, not financial terms. This analysis suggests an economic cost equivalent of around 40 units/kWh for diesel-drive pumps. If this value is supposed to set an upper limit on the true value of electricity used for pumping, then a value of 35 units/kWh, midway between this and the proposed tariff, seems a realistic figure to use.

Using a valuation of 35 units/kWh for electricity used for irrigation and industry, and 50 units/kWh for electricity for domestic and commercial use, gives an economic rate of return of around 9.7%. However, this return is based on a reasonably conservative valuation of the benefits of irrigation pumping, and ignores the various unquantifiable benefits associated with electrification of the towns, villages and farms in the area. Taking these into account as well, it is likely that the benefits of the scheme are greater than the costs, at a discount rate of 12%, and therefore that the scheme should go ahead.

Table 11.4   Economic return for different measures of benefits: Cash
flows

	Gross benefits of electrification				Net benefits of electrification			
	A	C	D	E	A	C	D	E
	DC:55	DC:50	DC:50	DC:55	DC:55	DC:50	DC:50	DC:55
	II:30	II:35	II:40	II:35	II:30	II:35	II:40	II:35
1978					-1933	-1933	-1933	-1933
1979					-4054	-4054	-4054	-4054
1980	361	400	446	407	-1749	-1710	-1664	-1703
1981	1012	1120	1248	1140	-1274	-1166	-1038	-1146
1982	1154	1280	1427	1302	362	488	635	510
1983	1297	1438	1604	1463	- 171	- 30	136	- 5
1984	1516	1683	1877	1712	- 434	- 267	- 73	- 238
1985	1710	1901	2122	1932	790	981	1202	1012
1986	1880	2091	2335	2125	1007	1218	1462	1252
1987	1931	2139	2384	2177	1132	1340	1585	1378
1988	1987	2192	2438	2234	1237	1442	1688	1484
1989	2056	2256	2503	2304	1290	1490	1737	1538
1990	2126	2322	2571	2375	1293	1489	1738	1542
1991	2206	2396	2646	2456	705	895	1145	955
1992	2293	2479	2730	2546	1425	1611	1862	1678
1993	2396	2575	2828	2650	1613	1792	2045	1867
1994	2507	2678	2933	2762	1706	1877	2132	1961
1995	2631	2794	3051	2888	1810	1973	2230	2067
1996	2631	2794	3051	2888	1810	1973	2230	2067
1997	2631	2794	3051	2888	1810	1973	2230	2067
1998	2631	2794	3051	2888	1810	1973	2230	2067
1999	2631	2794	3051	2888	1810	1973	2230	2067
2000	2631	2794	3051	9870	8792	8955	9212	9049

DC:55 : Domestic and commercial electricity supplies valued at 55
units/kWh

II:30 : Irrigation and industrial electricity supplies valued at 30
units/kWh etc.

Source: BERRIE, T.W.: 'Rural electrification'. Paper given to the
International Symposium on Electricity Economics and Load Management, Imperial College, London, 24-28 March 1980.   Reproduced by
permission of Imperial College, Power Systems Group

Table 11.5   Calculation of internal economic return

	Capital expenditure	Fuel cost	Salaries generation	Other costs generation	Maintenance transmission	Income at recommended tariffs (benefits)	Net benefits
1978	1933						−1933
1979	4054						−4054
1980	1874	86	29	70	51	354	−1756
1981	1878	242	29	70	67	992	−1294
1982	362	278	29	70	73	1132	320
1983	969	312	29	70	88	1272	− 196
1984	1386	365	29	70	100	1488	− 462
1985	286	413	29	88	104	1679	759
1986	193	455	29	88	108	1846	973
1987	109	463	29	88	110	1893	1094
1988	50	472	29	88	111	1945	1195
1989	55	482	29	88	112	2008	1242
1990	110	494	29	88	112	2073	1240
1991	748	506	29	106	112	2146	663
1992	100	521	29	106	112	2227	1377
1993		536	29	106	112	2321	1556
1994		554	29	106	112	2423	1640
1995		574	29	106	112	2537	1734
1996		574	29	106	112	2537	1734
1997		574	29	106	112	2537	1734
1998		574	29	106	112	2537	1734
1999		574	29	106	112	2537	1734
2000	6982	574	29	106	112	2537	8716

Source: BERRIE, T.W., 'Rural electrification'. Paper given to the International Symposium on Electricity and Load Management, Imperial College, London, 24–28 March 1980. Reproduced by permission of Imperial College, Power Systems Group

Table 11.6 Economic return for various measures of benefit: Summary

			Economic rate of return %
A	Domestic and commercial	55 units/kWh	8.3
	Irrigation and industrial	30 units/kWh	
B	Domestic and commercial	50 units/kWh	8.0
	Irrigation and industrial	30 units/kWh	
C	Domestic and commercial	50 units/kWh	9.7
	Irrigation and industrial	35 units/kWh	
D	Domestic and commercial	50 units/kWh	11.3
	Irrigation and industrial	40 units/kWh	
E	Domestic and commercial	55 units/kWh	10.0
	Irrigation and industrial	35 units/kWh	

Source: BERRIE, T.W.: 'Rural electrification'. Paper given to the International Symposium on Electricity Economics and Load Management, Imperial College, London, 24-28 March 1980. Reproduced by permission of Imperial College, Power Systems Group

Table 11.7 Costs of diesel-drive alternative

	Running costs* units/kWh	Capital costs units x 10^6 38.5kW	Capital costs units x 10^6 20kW	Annual charge† units/kWh	Total annual costs units/kWh
1980	47.8	348.4	1078.1	10.5	58.3
1981	32.4	–	269.5	9.3	41.7
1982	32.4	–	269.5	9.2	41.6
1983	31.6	–	269.5	9.3	40.9
1984	30.6	51.0	321.3	9.3	39.9
1985	30.0	51.0	321.3	9.1	39.1

* Units per kWh equivalent.
† Calculated as a 9.5% annuity over the 15 year life of the equipment.

Source: BERRIE, T.W.: 'Rural electrification'. Paper given to the International Symposium on Electricity Economics and Load Management, Imperial College, London, 24-28 March 1980. Reproduced by permission of Imperial College, Power Systems Group

REFERENCES

[1] 'Energy needs, uses and resources in developing countries'. Publication 50784, Brookhaven National Laboratory, Upton, New York, March 1979

[2] MUNASINGHE, M., and WARREN, C.J.: 'Rural electrification, energy economics and national policy in developing countries' (World Bank, 1979)

[3] World Bank Atlas, 1981

[4] MUNASINGHE, M., and WARREN, C.J.: 'Rural electrification, energy economics and national policy in developing countries' (World Bank, 1979)

[5] 'Small scale energy activities in India and Bangladesh'. Trip Reports No. MIT-77-02WP, MIT Energy Laboratory, Cambridge, Mass., USA, 31 Aug. 1977

[6] 'North-south: A programme for survival'. Report of the Independent Commission of International Development Issues under the Chairmanship of Willy Brandt (Pan Books, 1980)

[7] 'Partners in development'. Report of the Commission on International Development. Lester B. Pearson, Chairman. 1967

[8] HOWE, J.W., BEVER, J., KNOWLAND, W., and TARRANT, J.: 'Energy for developing countries'. Overseas Development Council, Washington DC, USA, 1978

[9] WARFORD, G., and SAUNDERS, R.: 'Village water supplies'. World Bank publication, Washington DC, USA, 1977

[10] 'Rural electrification'. World Bank Sector Paper, 1975

[11] World Development Report 1981. World Bank, Aug. 1981

[12] 'Standards of rural electrification'. Ref. 671-86; and 'Pricing of indigenous energy resources'. Ref. No. 672-15. World Bank research program, Abstract of current studies, Oct. 1980

[13] Other rural electrification guidelines of a later date than those of the World Bank are in 'Power sector planning manual' (UK Overseas Development Administration, June 1979) Chap. 8; also some booklets available from the National Rural Electrification Cooperatives Association (NRECA), Washington DC, USA

[14] The Appendices are based on: BERRIE, T.W.: 'Rural electrification study: Work programme - economics' and not on any particular study. The paper was given to the International Symposium on Electricity Economics ad Load Management, Imperial College, London, 24-28 March 1980. Reproduced by permission of Imperial College, Power Systems Group

Chapter 12

Application to developing countries

12.1 BACKGROUND

12.1.1 Introduction

Power-system economics in some respects requires a special application to developing countries, particularly those in the lowest per capita income group.[1] We have already touched on this in Chapter 11. In this Chapter we mention some of the other ways in which developing countries differ from developed in this respect.

Most of the extensive literature on power-system economics in developing countries comes from lending agencies like the World Bank and the UK ODA.[2]

12.1.2 Main reasons for the difference

We briefly list the most important reasons for difference below.

THE MAIN REASONS FOR DIFFERENCE BETWEEN
DEVELOPED AND DEVELOPING COUNTRIES

Firstly: There is extreme shortage of capital in developing countries

This means that both foreign and local capital are usually at a premium. If developing countries are to avoid crippling debt service charges they must obtain as much as they can of 'official' aid, e.g. from lending agencies such as the World Bank, rather than from Commercial Banks. However, these lending agencies apply strict technical, economic, financial and institutional rules in the interest of getting the 'best' project. The power utility must, among other things, become fully conversant with power system economics.

Secondly: Electricity supply is extremely capital intensive

A special watch must be kept on capital investment to check, for example, that:

(i)   standards of security are not excessive

(ii)   'appropriate' technology is used

(iii)   special steps are taken to avoid cost-overruns

(iv)   ex post-evaluation is carried out on all projects to determine
        the best way to avoid difficulties in the future.

Thirdly: Fuels can be the major issue in the power sector

In countries with large quantities of indigenous fuels, whether
already developed or not, whether inexhaustible or not, the wise use
of these fuels may be the main issue in planning all energy sectors.
For example, questions arise as to:

(i)   whether the indigenous fuels should be exported rather than
        used at home

(ii)   what the internal allocation and pricing of these fuels should
        be between sectors in the economy

(iii)   how quickly such fuels should be depleted

(iv)   what is an accurate reflection of the true 'resource cost' of
        these fuels to the economy.

In countries with no large quantities of indigenous fuels we must
take a view on:

(i)   the likely long-term comparative prices of alternative fuels
        which can be imported, e.g. coal, oil, or natural gas

(ii)   what fuels are being used by industry and the power sector at
        present

(iii)   what fuel should be used for what purposes in the future.

Fourthly: An optimum energy pricing policy is vital

The country's energy sector pricing policy must be set as near to
optimum in the economy as possible. As we saw in Chapter 9, it is
only by this means that the right 'signal' can be given to the
consumers of any fuel. In developing countries determining such an
energy policy is not easy, usually needing:

(i)   large inputs of data from all sectors of the economy

(ii)   models of the macro-economic sector, the energy sector and the other sectors of the economy

(iii) a large digital computer

(iv) well trained and experienced professional staff

(v) capital resources to carry out the studies.

The World Bank has adopted a policy of providing technical experts and loan finance for carrying out energy-sector reviews[3] especially in the poorer developing countries.

Fifthly: The economic methodology needs changing

The standard methods of project economic comparison used today are beginning to seem inappropriate for developing countries.

(i) The standard approach is to compare projects using some discounted cash flow method. Yet, if the main criterion for optimisation is, say, to avoid as much maintenance as possible, then why not substitute this as the criterion, or at least re-optimise the development programme using this as a very basic constraint? All practitioners in developing countries will recognise other serious similar constraints in this category.

(ii) In the past, mainly to impose capital rationing, high discount rates have been used to choose least-cost solutions, even though all economic sums are done in constant price levels, i.e. allowing only for relative price changes. Thus a commonly used discount rate of 12% implies being able to make a return of at least 30% in money terms. Furthermore, post-evaluation studies show little evidence that high discount rates choose the 'right' project in developing countries.[4] Capital-intensive projects are possibly being excluded to an un-warranted extent by using a high discount rate. Surely capital rationing can be done more effectively by other means? On balance it seems that, if discounted cash flow techniques should still be used, the discount rate should be lowered.[5]

12.2 MAIN CONSIDERATIONS[6]

12.2.1 The economy

It is particularly important in developing countries to see the subject of power-system economics within the context of the national economy, because, as we have seen, the power sector makes a large demand upon scarce local capital resources, is the dominant sector within the energy sectors, and uses up a lot of foreign exchange on capital works and fuel.

Therefore, we must view the power sector with all its requirements for the future within the development of the overall economy. If the power-system development plan is part of a national plan, then its rationale will automatically be correctly considered. Again, any national plan will give the power sector a lot of past, present and projected data, e.g. fuel prices, labour costs, foreign exchange rates, shadow prices of inputs and outputs which are more important in developing country planning than in a developed country.

12.2.2 The power sector

Developing countries vary considerably in their administration, technical development, size of total investment programme and number of identified projects in each sector. Each factor has a direct impact on the application of power-system economics; how the power sector is administered; its technical development; the sector's share of the national investment plan; how identified projects are fitted into the sector plan; the current basic problems of the sector and the recommended solutions.

Studies of the country's overall power sector are always important, and in some cases meaningful power-system economics cannot be done until such a sector review has been carried out, often together with some pre-investment studies for future projects.

The actual organisation and administration of the sector is important, particularly those utilities with special problems, e.g. on tariffs, financing, operational efficiency. We must also consider any national regulatory bodies on tariffs, capital investment criteria and institutional controls.

The technical capability of the sector in relation to its needs must be known, as it is the jumping-off place for applying any system economics. Present and foreseeable problems of quantity and quality of electricity service are always important.

If a power-sector development programme planned at the national level exists, the application of power sector economics will be that much easier. A broad judgment must be made about this plan and how it affects any particular utility and project. If the power sector is not controlled at any national level and/or there is no national programme for the development of the sector, then any meaningful application of power-system economics might prove difficult; some 'guestimates' of basic data mentioned earlier, however, must still be made.

12.2.3 Justification of projects[7]

The economic justification of a project answers the same question in developing and developed countries: is the project the most economic of all alternatives, including not investing in anything at all? However, the economic justification process in a developing country must of necessity be more intense; it must persuade a Finance Minister, besieged for resources by every sector, that the extension to electricity services takes preference over many transport, health, education and agricultural projects.

12.2.4 Finance

Some information on the finance of the utility and the national accounts is needed to apply system economics to developing countries; e.g. will the application of marginal cost pricing enable the utility to meet the financial target set by government; can the financing plan for the utility be fitted into the national plan; will any tariff increases be needed to meet the financial target and the financing plan?

The utility's present position and financial history enables a conventional financial analysis to be carried out of its accounting books.   If records are inadequate, then judgments must be made.

The utility's financing plan must contain not only the requirements and sources of finance for a particular project, but also that for any other work to be carried out by the utility over the same period. The degree of detail necessary will vary enormously and no rules can be laid down. Specific points will need specific attention before a plan can be accepted, e.g. how to deal with any large financing gap.

Important details from the three normal commercial financial statements, balance sheet, sources and application of funds, and the income statement, may be needed before any power system economics can be applied.

REFERENCES

[1]   1981 World Bank Atlas, Washington DC, USA

[2]   'Electricity sector paper' (World Bank, 1972); also 'Power sector planning manual' (UK Overseas Development Administration, 1979)

[3]   'Accelerating projects for the exploration of petroleum in developing countries' (World Bank, Jan. 1979); also 'Energy in developing countries' (World Bank, Aug. 1980)

[4]   Annual Performance Review of Post Project Audits, 1978, 1979, 1980, 1981. (World Bank, Washington DC, USA)

[5]   But there are other considerations, see Reference[2]

[6]   World Bank Staff Working Paper 350, Washington DC, USA. 1979

[7]   BERRIE, T.W.: 'Application of electricity economics and load management to developing countries'. Paper given at International Symposium on Electricity Economics and Load Management, Imperial College, London, 20-24 March 1980

Risk analysis in project economics

Introduction

Although what follows has been written for an agricultural project, the arguments used and the methodology are directly applicable to power-sector projects. Moreover, it is the most succinct and clear explanation of how to carry out risk analysis in connection with project justification known to the author.

It is reproduced by permission of the Economic Development Institute of the World Bank, Washington DC, USA.

Illustration of dealing with risk and uncertainty in project economic analysis[1]: Agricultural development project (The Gambia): Risk-analysis supplement

Based on our assessment of the <u>most likely outcome</u> of the project's rice component, the internal economic return (IER) is 34%.

This result is based on our assessment of the most likely outcome of a large number of events, such as: cost of project administrators, housing, pumps, earth-moving equipment; timing (relative to other parts of the project) of finding villages willing to cooperate, finding appropriate sites, hiring project staff, obtaining procurement and delivery of necessary inputs, getting villagers to level and bund the land; and yields per acre and farm price of the output - rice. <u>All these assessments</u> - the assumptions which went into the calculation that generated the 34% IER - <u>are less than certain</u>, just as all statements about the empirical world, and particularly about the future, are uncertain.[2]

But if results are other than those we assumed as the most likely prospects, then the economic (social) benefits from the project are most likely to be different from what the 34% IER tells us. <u>What is the project's rice component's internal economic return when the uncertainty of our assumptions is taken into account? How risky is the project? And more specifically, what are the chances that its return will fall far enough below the most likely IER so that it will cease to be socially worthwhile?</u>

Sensitivity analysis

One way to take uncertainty into account is called sensitivity analysis. This method is crude, but it is also simple, and, in many cases, it is good enough. Sensitivity analysis consists of:

1. looking at the major assumptions of the project analysis

2. picking out a few where you think the most uncertainty is involved

3. picking one or maybe a few other possible outcomes that seem plausible

4. adjusting project costs and benefits to reflect what would happen if this alternative outcome were to happen

5. calculating the new internal economic return

6. weighing mentally how likely you think the occurrence of the alternative outcome is; and how much of a disaster (or windfall) its occurence would be to the economy (society), as reflected in the new internal economic return.

In the Gambian exercise we adjusted the project's costs and benefits to account for six possible changes in the analyst's assumptions about the most likely outcome. Each of these alternative outcomes would have made the project less economically (socially) attractive than before. The sensitivity analyses gave IERs ranging from 18% to 71% lower than the original analysis.

In only one case - the assumption of constant inflation in cost vis-a-vis benefits (C) - did the 'sensitivity' calculation bring the project's return down so low that planners might consider abandoning it. The occurrence of that particular outcome was judged to be highly unlikely.

So sensitivity analysis, crude and simple though it is, was sufficient to show that the project was worth doing. Or was it? Suppose that some combination of the circumstances tested in the sensitivity analysis had occurred. Or suppose that the Gambia were really an extremely resource-short country so that opportunity costs of resources were in the 25-30% range (or that the project had been less dramatically attractive to start with). To be a little bit surer about our analysis, we would have to go beyond sensitivity testing.

The principal shortcomings of sensitivity analysis are that it is incomplete and ambiguous. It is incomplete because it considers only a small number of outcomes other than the most likely one, even though the ones it considers are judged to have the higher probabilities among the many less-than-most-likely possibilities. Also, sensitivity analysis considers things that could turn out in ways other than the most-likely way one-at-a-time. Actually, they might occur in groups, or they might be dependent on each other in complicated ways.

Sensitivity analysis is ambiguous in that it does not specify how likely or unlikely the occurrence of the alternatives tested is. At most, an implicit guess is made about the probability of each alternative.

Risk analysis is an extension of sensitivity analysis. To put it another way, sensitivity analysis is a special case of risk analysis. In risk analysis, the incompleteness and the ambiguities of sensitivity analysis are cleared up. But clearing them up may be difficult, and it takes time.

In risk analysis

1. we try to guess the probability that alternatives to our assumed most-likely outcome will happen

2. then we figure out what the internal economic (social) return would be for each combination of circumstances

3. we weight those rates of return to account for the likelihood that any combination will occur; and hence

4. we can come up with an adjusted IER that takes account of all the possibilities - at least, of all the ones we bothered to put into the analysis, and weighted by the probabilities we chose to give them. This is called the expected outcome or the certainty equivalent. Moreover,

5. we can then construct a cumulative probability distribution, which will show the probability that the IER will fall below (or rise above) any given point, and

6. we can calculate some summary statistic or riskiness, such as standard deviation of the series of possible internal rates of return.

The primary practical difficulty of doing risk analysis lies chiefly in its tediousness. Just to do a risk analysis with the six alternatives we tested in sensitivity analysis above, we would have to construct the net-benefit streams for each of the 64 combinations and do an IER calculation for each. If, instead of considering only one alternative to our most-likely outcome assumption in each of the six cases, we wished to include four alternatives, then the number of net-benefit streams to be constructed and of IER to be calculated would rise to 15,625.

Clearly, if you are going to carry out the risk analysis 'by hand', it is important to keep it very simple. It has to be kept so simple, in fact, that it is usually not very informative.

The alternative is to turn to electronic computers to speed up operations. A number of computer routines that are available will permit the consideration of a very large number of different possibilities. They do this not by calculating each one, but by taking a sample of all the different possibilities and basing their risk calculations on that sample. Since the number of possibilities is very large, the sampling technique approximates to the result you would get by calculating all the different possibilities. Even so, the sample is generally quite large, and it is necessary to specify the IER associated with each possibility in the sample - a task of formidable dimensions unless the various assumptions in the project analysis which affect the cash flow can be related to each other mathematically. That, in itself, can also be a formidable task, since interdependencies may be far from clear.

We will say no more here about how computers or computer-based risk analysis programs work.[3] For most analysers of projects, computers and the appropriate programs may not be available at the right time, place and cost. In any case, the principle of risk analysis is the same, whether done by hand or computer. The Gambian example that follows will serve to illustrate the principle.

The primary theoretical difficulty in risk analysis concerns dependencies among the various possibilities considered. Once we have completed all the steps mentioned above (e.g. spelling out all the possibilities and attaching weights to them), our problem is deterministic. In other words, we have made enough assumptions to determine the outcome, and the result is merely a matter of laborious calculation.

Our calculation assumes that the different parameters considered are independent. In fact, they never are totally independent; no two events are ever totally independent. It is easy to see that rainfall and yields are highly dependent in rainfed agriculture. It is less easy to see that cost and benefits are interdependent. There is no completely satisfactory way of dealing with dependencies.

The only practical advice is to do your best to reduce your risk analysis to a few parameters that are as mutually independent as possible. Risk analysis is based on the assumption that the parameters are independent. Unless that assumption is realistic, the analysis will be a waste of time and will produce meaningless results.

Risk analysis: A simple Gambian example

To illustrate the principle, without getting into the complications of computer operations, we look at the Gambian example.

The six alternatives we examined in sensitivity analysis are not all examples of uncertainty. The assumption about the value of family labour (B) - that is, whether the economic price is zero or half the official wage rate - is not at the mercy of uncertain future events that are beyond the control of the project management. There is something different between the question of whether the world price of rice or pumps will go up or down. There is a right answer to the former – the wage questions – only we are not sure what it is. For the latter – the prices – there is genuine uncertainty. Therefore, let us assume that our 'most likely outcome' assumption for valuation of family labour (that is, zero) is right, and drop the matter.

The remaining five sensitivity tests are examples of uncertainty. However, the assumption of constantly rising cost relative to benefit prices (C) is exceedingly unlikely. Therefore, let us forget about it.

Of the remaining four, the assumption about a year's delay in benefits (relative to costs, of course) (E) might be related to the other three, particularly to the cost-over-run test, in complicated ways. For simplicity's sake, to avoid problems of interdependency, let us drop that too.

The remaining alternatives concern the <u>cost estimate</u> (A), the <u>benefit estimate</u> (D), and the possible early <u>termination</u> of the project (F). We will assume that the three events are going to be determined independently, though no two events are ever completely independent. There is bound to be <u>some</u> outcome for each measure; the project will have some cost, some benefit, and either will or will not end before the second set of pumps wears out. The result of all three is beyond the control of the project in large measure, hence uncertain.

Our sensitivity analysis above considered only the possibility that things would turn out <u>worse</u> than the most likely outcome. From reading the project analysis and from inspecting the project, it seems that farmers might get better yields than expected. Also, in view of world events that have taken place since the World Bank's projection of the world rice price, on which the estimate of project benefits was based, the valuation of benefits might have been too conservative. Therefore, let us introduce the possibility that project benefits will be <u>10% higher</u> than anticipated.

Based on our knowledge of the project, we now need to guess the <u>probability</u> for each of the alternatives. It is difficult to estimate probability, but the difficulty is not intrinsically any greater than a lot of other estimates about the future which you have to make in project analysis. Here are my guesses:

<u>Cost estimates</u>	most likely	0.6
	10% higher	0.4
		1.0
<u>Benefit estimate</u>	most likely	0.7
	10% higher	0.2
	10% lower	0.1
		1.0
<u>Termination</u>	most likely	0.7
	early (after 1 pump-life)	0.3
		1.0

Something must happen for each of the three, and they are assumed to be independent. Therefore, there are 3×2×2 or 12 possible outcomes. These are listed below.

The probability of each of the 12 outcomes is the product of the probabilities of each of its components. Thus, the probability of the most likely outcome for all three factors – the one used in the original analysis – equals 0.6 (probability of realising the most likely cost estimate) × 0.7 (probability of most likely termination) × 0.7 (probability of most likely benefit estimate), or 0.294. That means that there is less than one chance in three that things will actually turn out the way the analysis expected that they would, even though it is easily the most likely.

The probability of each of the 12 possible outcomes is given below.

Now, the costs and benefits have to be adjusted to give a net-benefit stream for each of 12 possible outcomes. The IER has to be found for each. This has been done. The resulting IERs are given on the following page. You will note that the IERs for the 1st, 2nd, 5th and 7th cases listed there had already been calculated in our sensitivity exercise above.

Expected outcome

By multiplying the IER for each possible outcome by the probability of its occurrence and adding the products, we get a new IER which takes account of the uncertainties we foresaw according to the probabilities we assigned to them. This is called the expected outcome (or certainty equivalent) IER. It is probability-weighted. In this case, it is 30%.

Of course, there is no way, even after the project, to know whether our list of possible outcomes was complete and whether our probability guesses were accurate. Only one of the possibilities will occur. That result either is something we anticipated or it is not (in which case our list of possibilities was not big enough). That is all we can tell.

Now the most likely outcome rate and the expected outcome rate may coincide. This will happen if each of the project estimates of most likely outcome is, so to speak, 'in the middle' of the range of possibilities. In other words, the possibility that yields, for instance, will be greater than the most likely outcome is exactly the same as the possibility that they will be lesser and, moreover, the possibility that they will be greater by each amount is the same as the possibility that they will be lesser by the same amount. In other words, if the probabilities for each estimate are symmetrically distributed around the most-likely outcome, then the most-likely outcome IER and the expected outcome IER will be the same.

These two rates are not the same for our guesses about the Gambia project. Expected outcome is lower than most-likely outcome. That is because we have estimated that the chances that events will turn out to be worse that the original project-analysis estimates are

Twelve possibilities

(1) Cost estimate	(2) Benefit estimate	(3) Termination	(4) Proba- bility (1×2×3)	(5) IER	(6) Weighted IER (4×5)
most likely (0.6)	most likely (0.7)	most likely (0.7)	0.294	34%	9.996
most likely (0.6)	most likely (0.7)	early (0.3)	0.126	28%	3.528
most likely (0.6)	10% higher (0.2)	most likely (0.7)	0.084	42%	3.528
most likely (0.6)	10% higher (0.2)	early (0.3)	0.036	38%	1.368
most likely (0.6)	10% lower (0.1)	most likely (0.7)	0.042	25%	1.050
most likely (0.6)	10% lower (0.1)	early (0.3)	0.018	18%	0.324
10% higher (0.4)	most likely (0.7)	most likely (0.7)	0.196	26%	5.096
10% higher (0.4)	most likely (0.7)	early (0.3)	0.084	19%	1.596
10% higher (0.4)	10% higher (0.2)	most likely (0.7)	0.056	34%	1.904
10% higher (0.4)	10% higher (0.2)	early (0.3)	0.024	28%	0.672
10% higher (0.4)	10% lower (0.1)	most likely (0.7)	0.028	18%	0.504
10% higher (0.4)	10% lower (0.1)	early (0.3)	0.012	9%	0.108
			1.000		29.674%
					Expected outcome IER

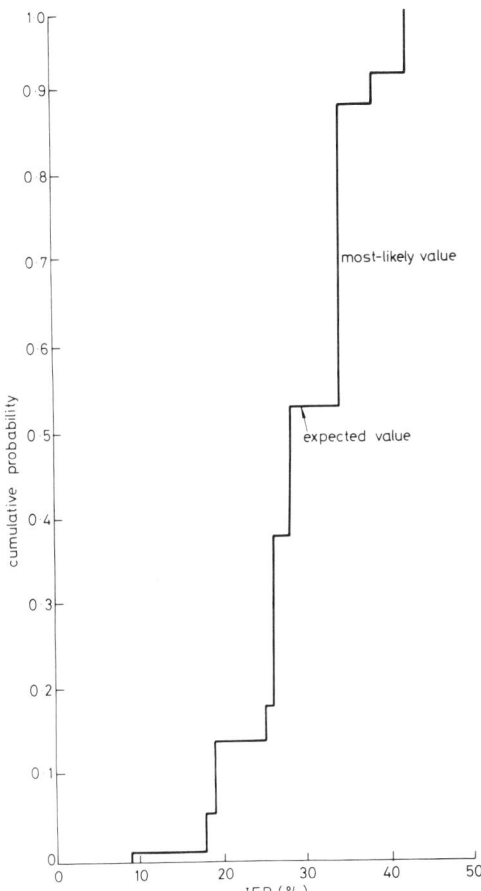

Fig. A1   Cumulative probability of IER

Source: Agricultural development project (the Gambia):
Risk analysis supplement, World Bank report. Reproduced
by permission of the Economic Development Institute of
the World Bank, Washington, D.C., USA

greater than the chances that they will turn out better.   Such an
assessment is probably reasonable, since project analysts generally
fall in love with the projects a little bit and, like all lovers,
become more optimistic than facts would justify.

Cumulative probability distribution

Unfortunately, whether expected outcome and most-likely outcome
are the same or whether they are not does not tell you how risky a
project is.   You can tell how risky it is, though, by using a simple
graph constructed from the table above. We simply rank the internal
rates of return which the various 12 possibilities would generate,
starting from zero, and graph their probabilities against the IERs
(see Fig. A1).

Possible IERs and associated probabilities

IER	Probability	Cumulative probability
9%	0.012	0.012
18%	0.046	0.058
19%	0.084	0.142
25%	0.042	0.184
26%	0.196	0.380
28%	0.150	0.530
34%	0.350	0.880
38%	0.036	0.916
42%	0.084	1.000
	1.000	

The distribution shows cumulative probability.   By examining it,
we can tell how risky the project is.
This distribution for The Gambia project reflects consideration
of a limited number of possibilities (12), so that the graph is a
broken line with fairly big 'steps'.   If a very large number of
possibilities were considered, and the probability of a large number
of possible rates of return were assessed, then the distribution would
look like a curve.
The distribution could be used in a number of ways.   Suppose that
the Gambian Government decided that the opportunity cost of resources
(OCR) is a certain percent, and we want to assess the chances that
the project will return less than that amount to the society
(economy).   The probabilities can be read from the distribution.

The illustration shows the practical use of the distribution. Note
that, in this case, there is a greater-than 50% chance that the IER
will fall below the most likely IER.   However, the likelihood that

Opportunity cost of resources	Probability that IER will fall below OCR
10%	0.012
15%	0.012
20%	0.142
25%	0.142
30%	0.530
35%	0.880

it will fall 'very much' below it (that is, by 0.25) is 'small' (less than 1/7). In practice, this is the most common way that risk analysis is actually used.

Standard deviation: A summary statistic

This practical example does not answer the general question, 'How risky is the project?' Only the whole cumulative probability distribution fully answers that question (under the assumptions we have made, of course). There is no way to explain the inherent riskiness of a project in one number. Nevertheless, there are summary statistics which can serve as a shorthand and which can give some indication of the riskiness of a project. One such summary statistic is variance. Variance is the probability-weighted sum of the squares of the differences between different possible IERs and the expected-outcome IER.[4]

Let us see by example just what that mouthful means. The expected IER is 30% (see above). The variance, then, is found as follows:

Difference between
expected IER and
different possible

IERs Squared		Probability		
$(42 - 30)^2$	×	0.084	=	12.1
$(38 - 30)^2$	×	0.036	=	2.3
$(34 - 30)^2$	×	0.350	=	5.6
$(30 - 28)^2$	×	0.150	=	0.6
$(30 - 26)^2$	×	0.196	=	3.1
$(30 - 25)^2$	×	0.042	=	1.0
$(30 - 19)^2$	×	0.084	=	10.2
$(30 - 18)^2$	×	0.046	=	6.6
$(30 - 9)^2$	×	0.012	=	5.3

Variance = 46.8

A more common summary statistic is standard deviation, which is the square root of the variance. The square root of 46.8 is 6.84; so the standard deviation is 6.8.

Do these summary statistics tell us anything we could not learn

from looking at the cumulative probability distribution? No. They
tell us less. The standard deviation of 6.8 tells us that there are
roughly two chances out of three that the IER will be within 6.8 of
the expected outcome, that is 30%.[5]   Stated differently, there is
about a two-third probability that the IER will be between 23.2% and
36.8%.

Conclusion

This supplement shows the way sensitivity analysis and risk
analysis work. Obviously, sensitivity analysis is a truncated and
subjective form of risk analysis. Risk analysis can be made much
more complicated than the example presented here. When it becomes
much more complicated than our example, you will certainly have to
use a computer routine, because hand calculations would just take
too long. The principle of the risk analysis does not change with
these complications, except that the computer constructs the cumulative probability distribution based on a sample of the possible
alternatives and not on all of them.

From the examples and discussion given here, you should be able
to go beyond sensitivity analysis to simple risk analysis in your
project analysis if you wish to. You should also be able to get a
fair idea of how much time and energy a risk analysis is likely to
take, when done 'by hand' and of whether you think that the result
is worth it.

It is my opinion that, for most development projects, sensitivity
analysis, intelligently used, gives a sufficient and efficient guide
to a project's riskiness. That was certainly true in the Gambian
case: our investment decision would not have been altered by the risk
analysis undertaken; it was not improved. In this particular case,
even the most sophisticated and extensive risk analysis would have
produced the same investment decision.

Not all projects are like this one, however. Where the most-
likely-outcome IER is close to the opportunity cost of resources (or
to the cut-off return rate for lending), and where there are
significant combinations of risks, and where we can make a fairly
convincing guess of the probabilities of the alternatives to our
most-likely assumptions, then simple risk analysis may change our
investment decision, and change it for the better. It may be worth
quite a lot of time and effort to get the decision just as right as
possible if large amounts of resources are involved.

Such circumstances are more the exception than the rule.
Unfortunately for risk analysis, the alterations that might change
our most-likely-outcome estimates are apt to be many and are apt to
be interdependent in complicated ways. Moreover, when the probabilities we assign to various events in risk analysis are highly risky
themselves, there is little likelihood that risk analysis will lead
to a better investment decision than some simple sensitivity tests.
And risk analysis takes more time.

REFERENCES

[1] Taken from some Course Notes of the Economic Development Institute at the World Bank in Washington DC, USA, and reproduced by permission of the Economic Development Institute. To obtain most benefit from the Notes, the power-system planner should assume that the phrase 'internal economic return' also covers the term 'equalising discount rate' between different alternative projects or development sequences (see Chapter 8).

[2] Economists have traditionally distinguished between risk and uncertainty. Both are events which may or may not occur in the future. For risky events, our knowledge of similar past events enables us to assign a probability to future occurrences in an 'objective' manner – the way life-insurance actuaries can assign a probability to your death at any given time. For uncertain events, we can only make a 'subjective' assessment of probable occurrence. In fact, the distinction between objectivity and subjectivity is very blurred. Consequently, the distinction between risk and uncertainty has been losing popularity with economists. In project analysis, we must assign probabilities to uncertain future events; we must do so as 'objectively' as possible. However, the past is usually a very uncertain guide to the future in such cases, because development projects are designed to alter radically existing conditions and because projects tend to be highly unique. Therefore, the events to which we are trying to assign probabilities, as objectively as possible, are nevertheless, very uncertain. Consequently, we shall ignore the risk-uncertainty distinction.

[3] Those interested in such computer applications might consult Shlomo Reutlinger: 'Techniques for project appraisal under uncertainty' World Bank Occasional Paper No. 10, 1970, Chapter 3. See bibliography for reference to the principal reviews of the vast literature on decision-making under uncertainty

[4] Note that we measure variance around the expected IER, 30%, which is the weighted mean of the series 9, 18, 25, 26, 28, 34, 38, 42. The simple mean of that series is 26.556, a number which does not interest us. For a similar example of this use of variance, see MAKEHAM, HALTER, and DILLON: 'Best-bet farm decisions'. Armidale, (Australia): University of New England, 1968, pp. 22-23

[5] If the distribution of possible values around the expected value were 'normal', then the proportion of possible values within one standard deviation of the expected value would be 0.6826; the proportion within two standard deviations would be 0.9546, etc. The normal distribution is the familiar bell-shaped (or

S-shaped) curve which is defined by the expected value, the standard deviation, and its formula. In our example, the distribution is not normal. It is not even symmetrical. In fact, if we look at the cumulative probability distribution, we can see that the probability that the IER will be within one standard deviation of 30% - that is, that it will be between 23.2% and 36.8% - is 0.738, not 0.6826. Using standard deviation for comparative purposes presumes that the distributions we are working with are normal, or nearly so. In the real world, the distribution of many characteristics is normal, or close to normal. However, it need not be so. In that case, standard deviation means only what its mathematical definition says it means, and no more.

Using discounting techniques
in project analysis

Introduction

The material in this addendum is based upon teaching material used by the Economic Development Institute (EDI) of the World Bank, Washington DC, USA, by permission of the EDI. Today there are readily available a large number of books on discounted cash flow (DCF) techniques, i.e. containing tables for discounting, compounding, present valuing, etc. The book which the author himself has found most helpful is: GITTINGER, J.P.(Ed.): 'Compounding and discounting tables for project evaluation', published on behalf of the World Bank by the Johns Hopkins University Press, Baltimore, USA, 1973. Table 1 showing the most commonly used numbers, i.e. the discount factors for 1.000, to three decimal places, is attached to this Addendum.

Compounding and discounting

The basic processes in DCF analysis (see Chapter 8) are those of compounding and discounting. For example, construction costs in constant price level terms are usually compounded up to the year of commissioning of the project, e.g. a power station, using an interest rate equal to the test rate of discount; running costs, including fuel, again in constant price level terms, are discounted from that same year and using the same interest rate. Although many books of tables show compounding and discounting figures to as many as six decimal places, or possibly 10 significant figures, bearing in mind the inherent inaccuracy of the data in most project work, only three decimal places are normally sufficient. Three decimal places will allow for estimating benefit to cost ratios to the nearest tenth of a unit ratio; and also for calculating net present values and internal rates of return to the nearest whole percentage point; both are as precise an estimate as is normally justified in a project.

Most compounding and discounting tables used for project work are computed assuming that the factors in the tables are used on the last day of the year for which they are stated. Working in shorter time intervals is not normally necessary. The point in time (t_0) to which compounding is done, or from which discounting is done, need not be the year in which the project is commissioned, but it usually is; it is sometimes chosen at another point of time which is convenient. To obtain 'present values' (see Chapter 8) of project costs, those values in a cash flow which arise before t_0 are increased to their equivalent value at t_0 through compounding, using the appropriate interest rate; those values which arise after t_0 are brought back to their equivalent value t_0 by discounting by the appropriate interest rate.

In project analysis, all elements in the cash-flow stream[1], i.e. all costs and benefits attributable to the project, assumed falling at the end of each year and including those incurred in the first project year t_1, are normally discounted to the beginning of the year of the commissioning of the project t_0. The basic formulas for compounding and discounting are:

Compounding factor for 1, i.e. $C = (1 + i)^n$, where:

C = present value, at the end of n years, resulting from the growth at a compound interest of i per annum, of an initial amount of 1.

Discounting factor for 1, i.e. $D = \dfrac{1}{(1 + i)^n} = \dfrac{1}{C}$, where:

D = present value, at the end of n years, resulting from the discounting of an initial amount of 1, by discount (interest) rate i per annum.

Discounted measures of project worth

In the manner described in Chapter 8 of this book, projects are normally ranked from an analysis of their attributable costs and benefits over their economic life; although in the case of ranking development programmes of projects, the period taken is longer, the principle applied is the same. Present values of cash flows are then found by compounding and discounting in the manner described above. Three discounted measures of project value are in common use:

$$\text{Benefit-to-cost ratio} = \frac{\text{Total present value of benefits}}{\text{Total present value of costs}}$$

Net present value = (Total present value of benefits)
                                      − (Total present value of costs)

Internal rate of return = Discount (interest) rate such that
                                      (Total present value of benefits)
                                      = (Total present value of costs)

These three measures, although they look different are essentially the same; i.e. for any project there exists an interest rate, which (a) is the internal rate of return, and (b) at which the benefit-to-cost ratio equals 1, and (c) at which the net present value is zero. The benefit-to-cost ratio is rarely used by the private sector or by public sectors like electric power; the internal rate of return is commonly used by governments and international lending agencies; in the power sector, the net present value is used primarily to determine the least-cost project from amongst a number of alternative projects which have the same benefits (see later).

Internal rate of return

Even when using a computer, the internal rate of return is usually determined by trial and error. This return measures the return, over the life of the project, to the investment and other resources attributable to the project. It should be noted in this connection that no assumption with respect to reinvestment need be made to be able to calculate this return.

To calculate the internal rate of return, first the cash flow of costs and benefits attributable to the project is discounted at some discounting rate believed to be near to the internal rate of return. By trial and error, a discount rate which is too low is found; i.e. when using this rate the net present value is positive. Also by trial and error, another discount rate is found which is too high, i.e. when using it the net present value is negative. The internal rate of return is somewhere between these two discount rates and we find it by interpolation:

$$\begin{array}{l}\text{Internal rate} \\ \text{of return}\end{array} = \begin{array}{l}\text{Lower} \\ \text{discount} \\ \text{rate}\end{array} + \begin{array}{l}\text{Difference between the} \\ \text{two discount rates}\end{array} \times (F)$$

where

$$F = \frac{\text{Present value of the cash flow at the lower discount rate}}{\begin{array}{c}\text{Absolute difference between the present values of the cash} \\ \text{flow streams at the two discount rates}\end{array}}$$

Note on depreciation allowances and interest

In DCF calculations, depreciation allowances in the accountancy sense are not included separately in the cost stream because DCF analysis itself automatically takes depreciation into account; i.e. if the undiscounted benefit-to-cost ratio is at least equal to 1, or if the undiscounted net present value is at least equal to 0, then the investment and other resources attributable to the project have been fully recovered. Thus, if the benefit-to-cost ratio at some positive discount rate is at least equal to 1, or if the net present value at some positive discount rate is equal to 0, or if the internal rate of return is positive, then, besides fully recovering the resources attributable to the project, there is an <u>additional</u> return on these resources equal to the internal rate of return.

As has already been seen, interest in the accountancy sense is automatically taken care of in the DCF process by compounding and discounting.

Least-cost solutions and equalising discount rates

As we saw in Chapter 8, finding least-cost solutions is one of the important tasks in power-system economics. It is used when we can be reasonably sure that the benefits of each alternative project or development programme are equal, e.g. when comparing alternative power stations with exactly the same loading pattern throughout their lives, or when comparing alternative power stations with different loading patterns but including the 'systems effects' (see Chapter 5). A common example of the latter is the comparison of a hydro power station versus a thermal power station (see Appendix 8.5); there are other common examples met with in designing networks and individual plant components, e.g. comparing transformers of the same rating, but which have different capital costs and running costs , the latter represented mainly by losses. In most of these cases an additional capital cost is being traded off against a smaller running cost (see later).

The ranking of alternative projects with the same benefits is done by listing them in ascending order of total present value of cost; the project which has the lowest total present value of cost is the least-cost solution. Thus the least-cost solution and the ranking may well change with discount rate. The most favoured least-cost solution subject to sensitivity analysis (see Addendum 1) is that using the opportunity cost of capital as the discount rate.

The discount rate which equates the total net present value of cost for two alternative projects, or development programmes, having identical benefits is often known as the equalising discount rate. In the common case mentioned above of trading additional capital cost against lower running cost, the equalising discount rate is a measure of the return on that additional capital.

Annualised values of cost

As we saw in Chapters 5 and 8, we often wish to replace a cost occurring at one single point in time by a level annualised cost which occurs in every year of the economic life of a project. There are two ways of approaching this, both giving identical results.

In the first method of approach, a 'sinking fund' type of depreciation is allotted to the capital cost. To find the annuity factor, i.e. the number by which to multiply the capital cost to get the annualised cost, the interest rate is then added to the sinking-fund factor. The sinking-fund factor is the level amount required each year to reach 1 by a given year at a compound interest rate equal to the rate of interest being used to find the annuity factor; this is normally the opportunity cost of capital or the test rate of discount.

In the second method of approach, a capital recovery factor is found that will repay a loan of 1 by a given year. The annualised cost is then found by multiplying the capital cost by the capital recovery factor.

Both sinking-fund factors and capital-recovery factors are found in the books of DCF tables referred to above.

We can see algebraically that the two approaches lead to the same end.

Annuity factor = F = Sinking fund factor + Interest rate

$$= \frac{i}{(1+i)^n - 1} + i$$

where

i = interest rate used

n = the number of years the sinking fund is to apply

$$\text{Thus} \quad F = i\left[\frac{1}{(1+i)^{n}-1} + 1\right] = i\left[\frac{1+(1+i)^{n}-1}{(1+i)^{n}-1}\right]$$

$$= i\left[\frac{(1+i)^n}{(1+i)^{n}-1}\right]$$

$$\text{Capital recovery factor} = R = \frac{i}{1-\text{Discount factor}}$$

$$= \frac{i}{1-1/(1+i)^n}$$

$$= i\left[\frac{(1+i)^n}{(1+i)^{n}-1}\right]$$

REFERENCE

[1]   BALDWIN, G.B.: 'Discounted cash flow techniques'. Finance and Development, Washington DC, Dec. 1977

Index

Benefits attributable to power system
projects 160-161
Black-outs *see* outage costs
Brown-outs *see* outage costs

Capacity
Constraints in system models 125, 127
Element in charges to consumer 178
Ideal marginal capacity-related costs
196, 198
Long-run 198-199
Pooling function of networks 210
Capital expenditure, *see* investment
Cash Flows
Comparison of alternative projects
163, 168-169
See also discounted cash flow (DCF)
techniques
Circuit lengths, in network configuration
synthesis 132-133
Coal
Proved reserves 8, 10
World data bank 8
World production estimates to 2020
19-20
Combined costs method 146-147
Compounding, in discounted cash flow
techniques 289-290
Conservation programmes
Effect on world energy demand
estimates 26-27
In developed countries 18-19
In supply and demand calculation 19
Constant-load-factor method, in choice of
power system 152-153
Consumers
Expectations of supply reliability 56
Initial reaction against rural schemes 249
Need to understand tariffs 40
Residential, effect of outages on 58-59
Services to, element in charges 178, 200
'Willingness-to-pay' factor in outage
cost estimates 56-57

Consumption
Effects of tariffs 38
Increasing through investment 35-36
See also demand, load forecasts control
devices, to limit consumption 187
Costs
Annualised values 292-293
Calcualtion 53-59
Estimating 56
Formal derivation for reliability rule
60-65
In assessing project viability 36
Of reliability 44-45
Outage 45-47

Decomposition methods, applied to network
synthesis 137
Demand
Evidence of suppressed demand 76
Extrapolation of historical trends 26-27
Likely position in 2020 28
Need for long-term projections 105-106
Peak constraints, in system modelling
122
See also consumption; energy require-
ments, world-wide; load forecasts;
spot pricing
Depreciation allowances, in discounted cash
flow 291-292
'Sinking fund' type 292-293
Developing countries
Application of power system economics
270-275
Choice of hydro-thermal systems 157
Fuels 2-3
From non-commercial sources 17
Importance of power system economics xi
Investment of power stations 31
Resource estimates 16-17
Rural electrification programmes
248-249
Total energy requirements 243
See also rural energy

Discounted cash flow (DCF) techniques 289
    Compounding and discounting 289-291
    Equalising discount rate 292
    Internal rate of return 291
    Sinking-fund and capital recovery
        factors 293
Distribution
    Comparison of systems 228
    Investment in 220-221
    Planning plant margin 228-229
    Radial systems 225
    Reliability 222
    Need for more case studies 229-241
    Ring systems 225-226
    System rehabilitation function 221-222
    System reinforcement function 221
    System replacement function 222

Econometric forecasting methods 80
Energy constraint, in system modelling 130
Energy-related costs, marginal 199-200
Energy requirements, world wide
    Overall situation 17-18
    supply-demand forecasts 19-20
Environmental factors, effect on tariffs 40
Equalising discount rate 168-169
Equipment, life expectancy and value 23
Exajoule (unit) 19, 26

Fissile fuels
    Breeder reactor effects on uranium
        demands 24
    Proved reserves 14-15
Forecasts, *see* econometric forecasting
    methods; load forecasts; market forecast-
    ing method
Frequency standards 215
Fuel cost savings 91-93
    In net effective cost method 153-154
    Merit order of alternative generating
        plants 92
Fuels
    Alternative sources 2
    Availability and price constraints 1, 86
    Factors influencing 'preferred' quality
        5, 8
    In developing countries 2-3
    Inexhaustible, characteristics 14-15
    See also geothermal power; hydro
        power; tidal power; wave power
    Money prices, leading to load forecast
        distortion 79
    Non-commercial
        Factors against use 248
        In rural areas 247
    Non-renewable
        In developing countries 16-17
        Reserve estimates 4-5

Resource forecasts 3-5
    See also coal; fissile fuels; petroleum
Renewable
    Energy output measurement 25-26
    Importance of future use 31
    World energy production estimates
        to 2020 26
Renewable but exhaustible 14
Substitution between dominant types 18

Generating plant, *see* power generation
Geographical consideration in transmission
    207-208
Geology, effect of changing circumstances 4
Geothermal power, assessment of potential
    15-16
Global models 140-146

Hydro power
    Marginal energy cost considerations
        199-200
    Network economics 209-210
    Plant, compared with thermal plant
        162-174
    Utilisation and potential 16
    World production estimates to 2020
        24-25
Hydro-thermal systems, choice of system
    157-159

Income redistribution
    Constraint on tariffs 40
    Objective of power projects 37
Integer programming 147
Interest, treatment in discounted cash flow
    techniques 292
Internal economic return (IER) on projects
    276
    Expected outcome (certainty equivalent)
        281
Investment
    Economic return
        Calculating 160-161
        For least-cost solutions 36
        On networks 212
        On rural electrification 260-264
        See also discounted cash flow
            techniques: internal rate of
            return; internal economic return
            (IER)
    Estimate for 1981-90 43
    In distribution 220-221
    Objectives 33-37
    Sum per consumer 70
    Tests of acceptability 31
    Trade-offs 34
Investment rule 34, 117
    And spot pricing 194-195

Least-cost solutions 211, 292-293
Linear programming, applied to power
   system planning 120-121, 140
   In global models 141-145
Load and plant duration curves 94-102,
   140, 197, 201-202
   In system models 123-124
Load factors, see constant-load-factor
   method
Load forecasts
   Ability to change 75
   And distribution growth 226-227
   Applications of modelling 104
   By national organisations 106-108
   Data collection 76
   Distorted by using money price of fuels
     79-80
   Economic, cost-benefit criteria 82-83
   For rural schemes, long-range 258
   In planning 69
   In rural areas 82-83, 254
   Macro-economic and econometric
     methods 80
   Market forecasting method 77-79
   Objectives 69-70
   Parameters 70
   Period covered 71, 77
   Practical application 73
   Principles 71-73
   Trend forecasting method 77
   Uncertainty 80-82
   Use of statistics 73, 75
   Using least-cost solutions 211
   See also consumption; demand
Load management 181-189
   By use of load limiters 187
   Centralised control 188-189
   Communication methods 189
   Control of peak usage 181-183
   Interactive control 189-195
   Use of differential tariffs 182-183, 187
Loss-of-energy expectation (LOEE) index 52
Loss-of-load probability (LDLP) index 52

Macro-economics, in load forecasting 80
Marginal analysis
   In generating plant planning cycle
     93-94
   Models 137-139
Marginal-cost pricing 39, 177-204
   For rural tariffs 256
   See also spot pricing
Marginal costs
   Long-run
     calculation 196-200
     of energy 199-200
     of transmission and distribution 199
     World Bank model 200-204
   Of rural electrification 252-254

   Use in choice of project 151
Marginal value, implicit, of power supplies
   46
Market forecasting method 77-79
Metering costs 40
Modelling 104-114
   Applications 113-114
   'Decomposition' process 118-120
   Details of model 110-114
   Existing models 105
   Feedback systems 119-120
   Methodology 108-109
   Power system economics, principles
     of 137-147
   Power systems 120-122
     Examples 121-132

Natural gas
   Proved resources 10-14
   World production estimates to 2020
     22-23
Net effective cost method 153-157
   Limits of applicability 156-157
Net Present cost calculations 36
Network expansion planning 131-132
Networks
   Capacity pooling function 210
   Configuration synthesis 132-137
   Continuous development 226-227
   Economic justification 211-212
   Group transfer criterion 133-135
   In England and Wales 218-219
   Major national 218-220
     Likely developments 219-220
   Miscellaneous functions 211
   Perturbation methods of solution
     135-137
   Planned transfer function 209-210
   Role as multi-purpose projects 210-211
   Role in economic operations 210
   Structure formation 208-209
   See also distribiton; transmission
Non-linear programming, in global models
   141, 145
Nuclear power generation 2, 18
   Fuel costs 155
   Possible position in 2020 28
   World production estimates to 2020
     23-24
   See also fissile fuels

Objective function, in system modelling 130
Opportunity costs 34
   In comparison of alternative projects
     162-163
Outage costs 45-47
   Calculation 53-60
   Consumers 'willingness to pay' 56-57
   Estimation 56

Formal derivation for reliability rule
   60-65
In system models 122-123
Of public illumination 59
To residential consumers 58-59
With improved reliability 49

Peak demand, inability meet 52
Perturbation methods 135-137
Petroleum
   Magnitude of price rises 1-2
   Proved reserves 9-14
   World production estimates to 2020
      22
   See also shale-oils; tar-sands
Power Generation
   Auto- and co-generation 90
   Calculating optimum mix 86-88, 93-94
   Elements in charges to consumers 178
   Formulating background plan 88-94
   Fuel cost savings 91-93
   Load and plant duration curves 94-102
   Marginal analysis 93-94
   Planning from present situation 91
   Plant reliability 51
   Plant to load balances 89-91
   Time-scale to station commissioning
      105-106
   See also network expansion planning
Power sector reviews 273
Power system economics
   Application to developing countries
      270-275
   Areas excluded 32
   Cost-benefit analysis 33
   Differences between developed and
      developing countries 270-272
   In network planning 132
   Modelling principles 137-147
   Subjects covered 33
   Use of engineers and economists 32-33
   Within total economic framework
      272-273
Power systems
   Attributable benefits 160-161
   Capital investment xii
   Choice, calculating economic return
      159-161
   Comparison of alternative projects
      162-174
   Finance in national context 274
   Historical overview 32
   Justification of projects in developing
      countries 273-274
   Load and plant duration curves 94-102
   Modelling 120-121
      Examples 121-132
   Multi-node 207-208

Planning
   Aims 117
   Choice of methods 151-157
   Choice of system, criteria 37
   Ranking of projects 292
   Supply quality, *see* reliability levels
Prices
   Factors causing distortion 40
   Factors in load forecasting 49-50, 72
   Related to demand 104
Pricing
   And threshold per capita income 79
   Power supply projects 38
   See also marginal-cost pricing
Pricing rule 38-40
Productive activity, cost of outages 57-58
Public illumination, cost of outages 59

Reliability
   In distribution 222
   Of substations 131
Reliability levels
   Costs incurred 44-45
   Indices 52
      Comparison 52-53
   Measures 51-52
   Of generating plant 51
   Optimum standards 43-44
      Defining, for system models 124
      Feedback 49-51
   Practical methodology 47, 49
   See also outage costs
Reliability rule 34, 43, 117
   Formal derivation of system costs and
      outage costs 60-65
Risk, distinguished from uncertainty 287
Risk analysis
   Computer applications 279, 286-287
   Cumulative probability distribution 284
   Example 279-282
   In project evaluation 276-288
   Principles 278
   Value of simplicity 278
Rural electrification 248-268
   And country's resource capabilities 255
   And development of tourism 254-255
   Basic ground rules 252
   Calculation of economic benefits 255
   Characteristics 250-251
   Economic aspects, methodology for study
      253-260
   Economic return, worked example
      260-264
   Economics of autogeneration 254
   Economics of connecting water pump,
      mills, etc 253-254
   Fuel costs and energy policy
      considerations 259-260
   Justification 251-252

Of individual projects within scheme
260
Preparing capital budgets 260
Problems for government and utility 251
Quality of supply 254
Sales and demand forecasts 258
Survey of consumers 257-258
Rural energy
Importance of fuel prices 243
Pattern of use 248
Quantity and quality 243-247
Sources and usage 246-247

Sensitivity analysis
Applications 277
In power system modelling 121
Limitations 277-278
Of cost parameters of alternative projects
168-169
Of load growth of alternative projects
174
Value 286
Shadow pricing 174
Shale-oils, world energy production estimates
26
Simulation models 139-140
Spot pricing 190
And investment rule 194-195
Principles 190-194
Standard deviation, use in risk analysis 285
Standard of living, influence of increased
consumption 35

Tar-sands, world energy production estimates
26
Tariffs
Accounting approach 180
Capacity-related charge 178

Commercial considerations 179
Components 177-178
Constraints 40
Determination of levels and structure 172
Divergence from marginal costs 39
Economist's view 180
Effects on consumption 38
Energy-related charges 178
For peak and off-peak periods 182-183,
187
For rural electrification 256-257
Methods of setting 38
Practical considerations 179
Strategy for setting 180-181
Time-related 39-40
Use of marginal-cost pricing 177
Technological developments
Effect on future costs 152
Effect on resource estimates 4
Thorium, *see* fissile fuels
Tidal power, suitable sites 15
'Tie-line' connections 187
Transmission, geographical considerations
212-220
See also networks
Transport factor in planning 89
Trend forecasting method 77

Uncertainty
Distinguished from risk 272
In load forecasting 80-82
Uranium, *see* fissile fuels

Variance, use in risk analysis 285

Wave power, assessment of potential 15
Wood fuel
Alchohol extraction for gas-oil mix 14
Proved reserves 14